I0060281

Let's Laugh at Trig!

Volume I: A Simple Introduction to Trigonometry

Kaveh Mozafari

ExcellenSation Inc.
370 Steeles Avenue West, Thornhill, Ontario, Canada, L4J6X1
Visit the author's website at **www.kavehmozafari.com**

Printed in Canada
First Edition 2016

ISBN: 978-09-940739-4-5

Table of Contents

Dedication

I would like to dedicate this book to my dad who tirelessly pursues his studies and gives me proof on a daily basis that it is never too late. Also, I would like to dedicate this book to my mother who knows nothing but kindness and sacrifice.

Preface

The fact is, it is easy. What is easy? It. Everything that is known. As long as you know it, it is easy, and if you have no knowledge about it, then you do not know about its level of difficulty. Even in the further case, we can safely assume it is easy as there is a fifty percent chance that it is easy. Taking into account that the existing acquired knowledge has shown to be clear, then I believe nothing prevents us from assuming otherwise. Regardless of the aforementioned points, there are at least two categories of information—the ones that we know that are easy and the ones we don't know. I believe if something is known (therefore, easy), it must be passed on in the simplest way possible. This method of handling knowledge leads to advancement towards the unknown. Otherwise, it is a time-consuming struggle to understand what is known. I also believe passing on information in the simplest way possible is a challenge that any wisdom lover (e.g., philosopher) must endeavour. The mentioned lines are what I firmly believe, and it is what I have kept in mind while writing this book.

Trigonometry is an ocean, and you can find it everywhere. It explains many, if not all, phenomena. Let me explain. If you explain anything, that means you have knowledge about it. That knowledge was acquired through an experiment or a logical thought process, which led to a conclusion. That fact means the process can repeat. That is, you can either experience it again or use the same logic to reach it. In simple words, it can be repeated. It can be periodic. Trigonometric functions and their variation can explain any routine event.

Let me be a bit more explicit. Let's say you are a morning person. As you wake up in the morning, you have a lot of energy. During lunchtime, you probably have lost some power, and you gain more as you eat food. While the day comes to an end, you have used the energy, and you go to sleep. This cycle repeats; therefore, it can be correctly explained by a sinusoidal function. You add more variables to it, and it looks even better. The question is, so what?

I leave that question unanswered. I want to give you another example. You are in an intimate relationship in which you experience ups and downs. You notice that the more you show how much you care, the more surprised you get from his/her behaviour, hopefully in a positive manner but possibly not in your favour or as per your expectations. What if you can write this periodic action as a function? What if you knew how much is too much? What if the brains of people in successful relationships are already creating these regular patterns to help them in their relationships?

I just wanted to present a few examples that you may have overlooked. A portion of this book (not in this volume) discusses such examples and applications. Now, the answer to the question (so what?): So you have a better chance than guessing to improve in many aspects of your life as a person, as a member of the society and as a creature on the planet Earth.

My intention in writing this book is to introduce ways that we can apply mathematics, or, specifically, trigonometry, in the simplest way possible to understand while providing numerous examples.

Acknowledgements

Almost any person I've interacted with has affected me while I was preparing this book. However, I would like to name a few people that I believe had a greater impact on this volume. The first people who come to mind are my caring parents, who taught me more than I can understand, my two sisters, Sahar and Sara, who have unconditionally supported me in all stages of my life, and the numerous people I have had the privilege of studying with. They had a great effect in the way that I wrote this book based on the feedback I received from them. A dear friend, Henry Le, who read through all my writings and without whom no one (not even I) would have been able to understand my sentences. I would like to take this opportunity and thank not only the aforementioned names but also everyone else who have been part of this book, which definitely would include you, the readers.

Introduction

"Let's Laugh at Trig" is a series of books that attempts to approach mathematical concepts of trigonometry in the most straightforward and understandable fashion. Employment of specific methods of writing intends to engage the readers using a friendly tone. Wherever possible, the concepts are mingled with historical data to provide a better vision towards the acquisition of the ideas.

The book assumes no mathematical background; therefore, it provides clarification for almost everything. Most of the information is intended to be appealing to readers of all backgrounds.

Most of the concepts in trigonometry can be explained in more than one way. This book has put forth a lot of effort to hand pick the easiest and most memorable ones. The reason lies in the belief that, after an understanding the basics, we can dive deeper into more complex concepts.

The current volume consists of six chapters, and it focuses on the foundations of trigonometry. This book provides numerous solved examples to equip students with more ways to practice and test their knowledge. There are also descriptive videos available on my website (www.kavehmozafari.com) that serve to further provide an explanation for various topics. Furthermore, the website contains a discussion page for interested people who have questions and those who would like to answer these issues.

I hope this book gives you a better understanding of the world.

Kaveh:D

1. Simple Approach to Trigonometric Nuts and Bolts

1.1. What are Sin(x) and Cos(x)?

I believe you expect definitions for sin(x) and cos(x). Let's put that aside and have some fun. Have you ever seen a normal coordinate system? If you haven't, the image below is an example of one.

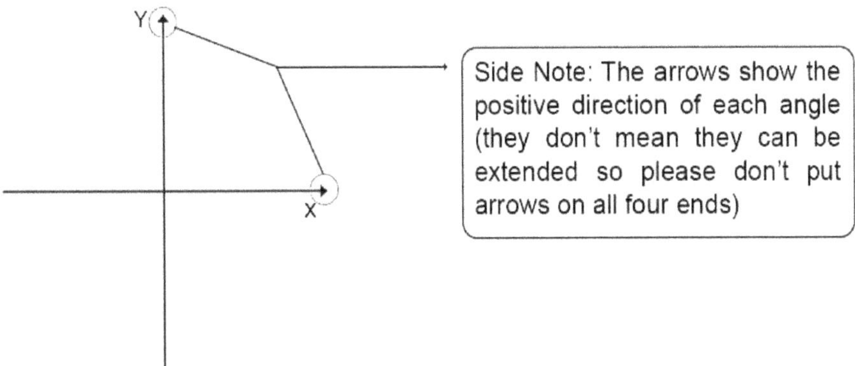

Figure 1.1 .Coordinate system

Now look at the X and Y axes. Are they moving? Of course not! They represent points on the horizontal and vertical axis, respectively. Consider the following graph:

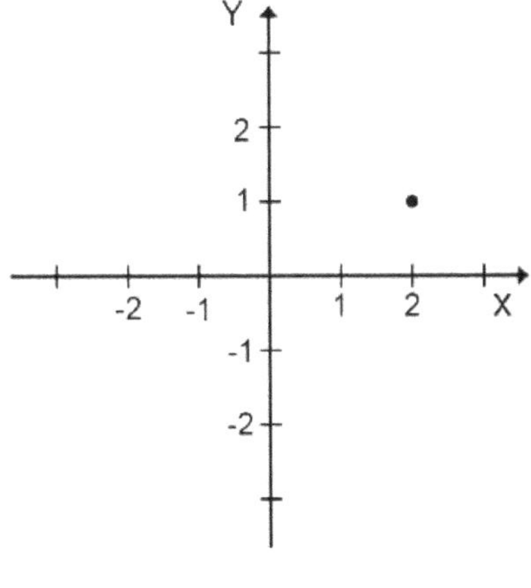

Figure 1.2

The above example represents point (2, 1), so the value of x is 2 and the value of y is 1. We can do this by drawing a straight line from the point on each axis. The line should be perpendicular to both axes. Otherwise, it would not show the correct point. Now try to find the coordinates of the following points. You will soon discover why they are critical. Please check and see if my answers are correct for the following examples:

Examples:

1.

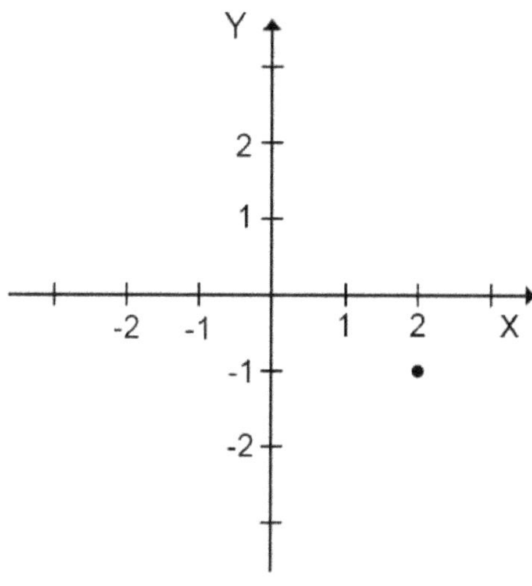

Figure 1.3.

Solution:

(2,-1) so the value of x is 2, and the value of y is -1.

2.

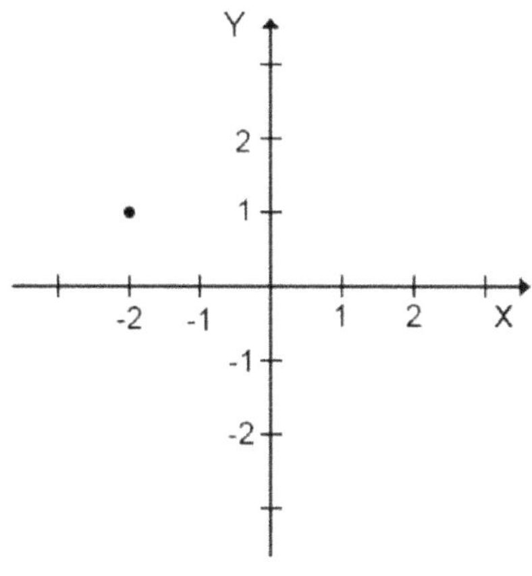

Figure 1.4.

Solution:

(-2,1) so the value of x is -2, and the value of y is 1.

3.

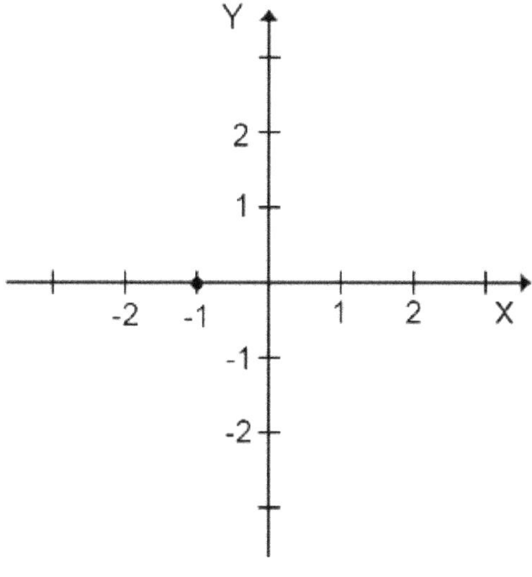

Figure 1.5.

Solution:

(-1,0) so the value of x is -1, and the value of y is 0.

4.

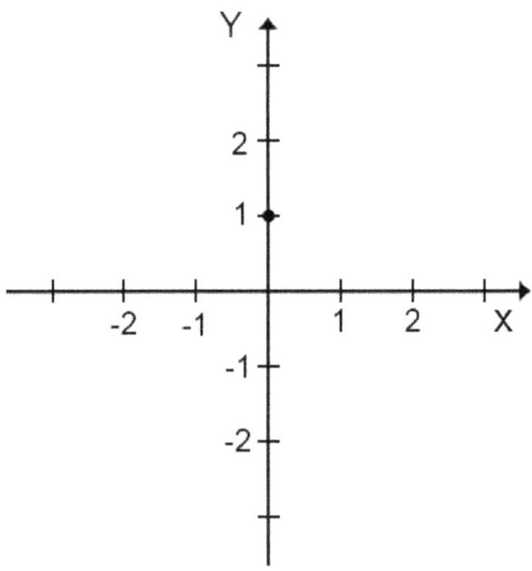

Figure 1.6.

Solution:

(0, 1) so the value of x is 0, and the value of y is 1.

5.

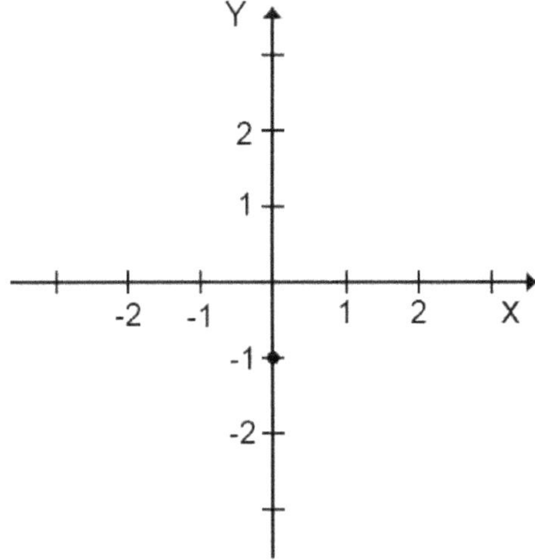

Figure 1.7

Solution:

(0,-1) so the value of x is 0, and the value of y is -1.

6.

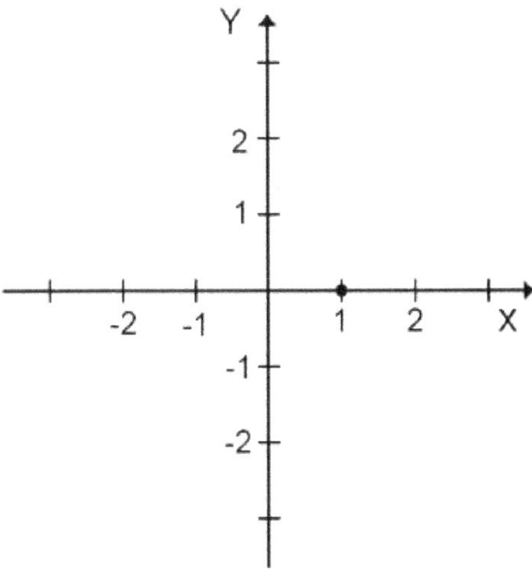

Figure 1.8.

Solution:

(1,0) so the value of x is 1, and the value of y is 0.

Now assume I would call y-axis sin(x) and x-axis cos(x). With the aforementioned assumption, we agree mathematicians may use the copyright laws against us. I assume Surya Siddhanta, who discovered the sine function, and Nasir Al-Din Al-Tusi, along with Al Battani, did not patent the information that they passed on to the Europeans. Based on our assumption, please find the coordinates and value(s) for sin(x) and cos(x) for each of the following variables:

Examples:

7.

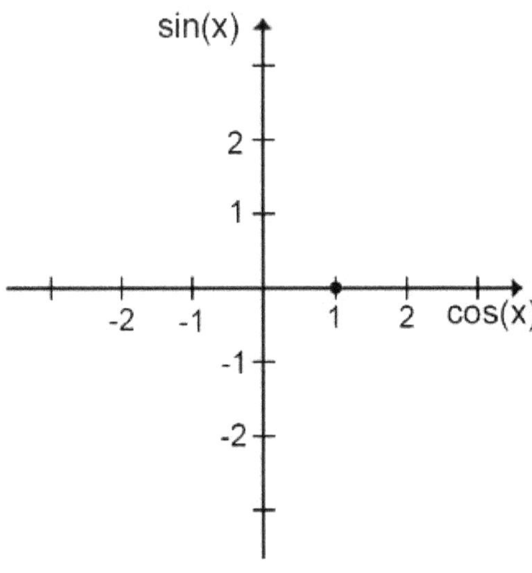

Figure 1.9.

Solution:

(1 ,0) so the value of cos(x) is 1, and the value of sin(x) is 0.

8.

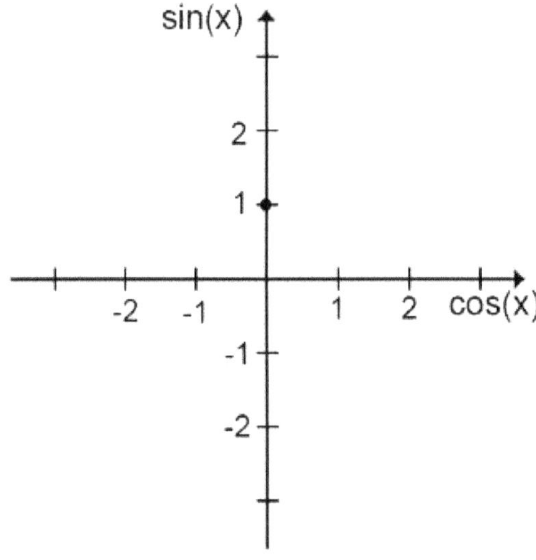

Figure 1.10.

Solution:

(0,1) so the value of cos(x) is 0, and the value of sin(x) is 1.

9.

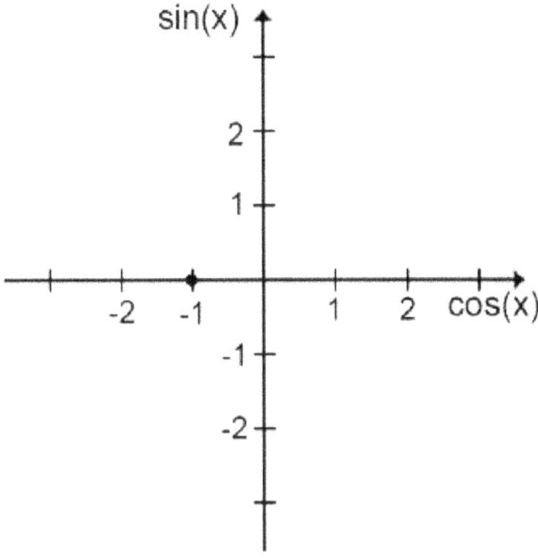

Figure 1.11.

Solution:

(-1,0) so the value of cos(x) is -1, and the value of sin(x) is 0.

10.

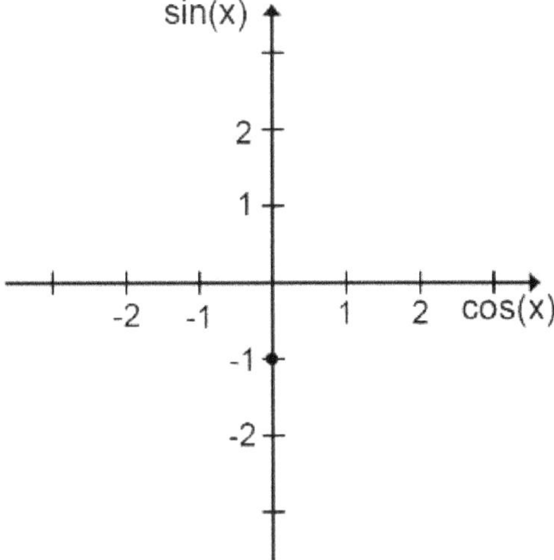

Figure 1.12.

Solution:

(0,-1) so the value of cos(x) is 0, and the value of sin(x) is -1.

Now it is time to invent a game. Let's place a circle with a radius of 1 on our coordinate system, in a way that its center is at point (0,0). The first rule of our game is that you are only allowed to pick points that lie on the circle. It means that the greatest value of sin(x) and cos(x) is 1. Try to find the coordinates of each point. Also specify the values of cos(x) and sin(x).

Examples:

11.

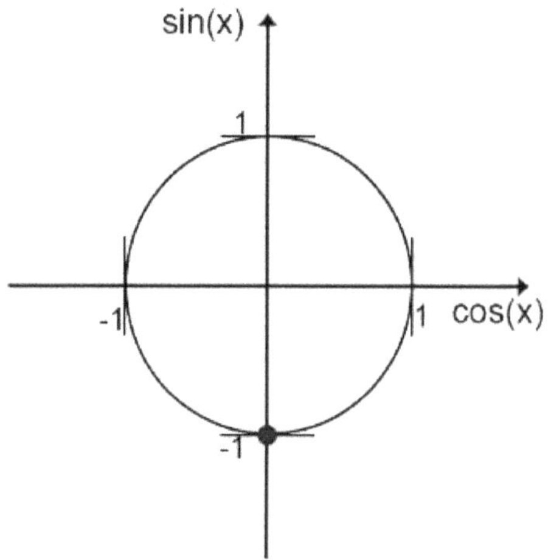

Figure 1.13.

Solution:

(0,-1) so the value of cos(x) is 0, and the value of sin(x) is -1.

12.

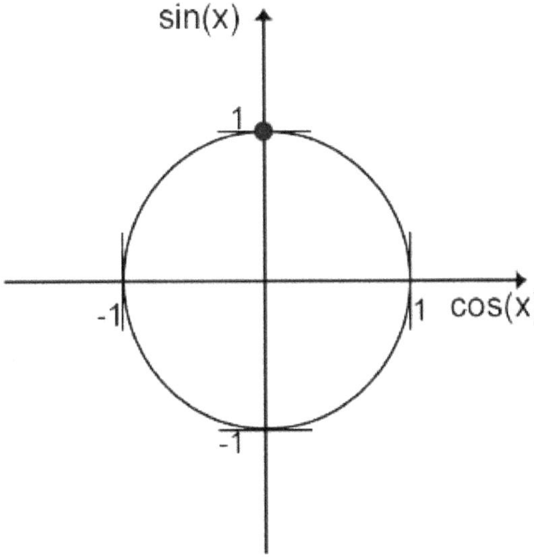

Figure 1.14.

Solution:

(0,1) so the value of cos(x) is 0, and the value of sin(x) is 1.

13.

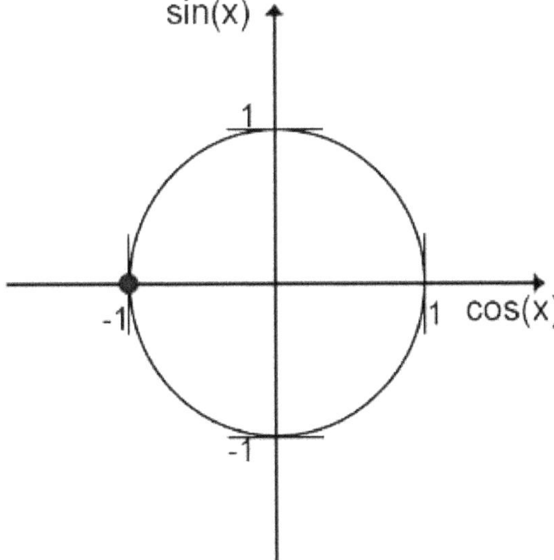

Figure 1.15.

Solution:

(-1,0) so the value of cos(x) is -1, and the value of sin(x) is 0.

14.

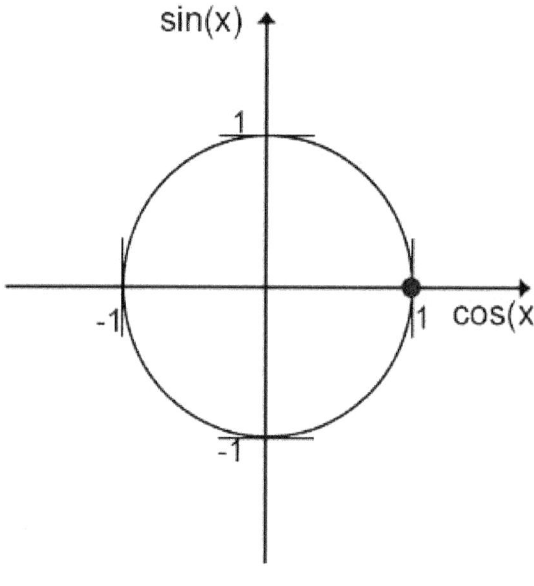

Figure 1.16.

Solution:

(1,0) so the value of cos(x) is 1, and the value of sin(x) is 0.

1.2. Distance Between Two Points:

We want to find the distance from point A to B, given we know how much we have to go left, right, up or down. For example, we find out if we are at point A, we need to go three units left and four units up to reach point B. In order to determine the distance, we need a measuring device such as a tape measure. Please use your tape measure to find the distance from point A to B (or length of line c). You don't have a tape measure? Many people from around the world have solved the problem without the use of a tape measure. Among those were the Babylonians by Pythagorean triples (2000 B.C.), the Chinese as mentioned in the book of "Zhou Bi Suan Jing," "Gougu Theorem" (202 BC to 220 AD), and Greek mathematician Pythagoras of Samos "Pythagorean theorem." The theorem is now best known as the Pythagorean Theorem (569–475 BC). If you really have a tape measure, don't embarrass yourself and use the theorem. The theorem is as follows:

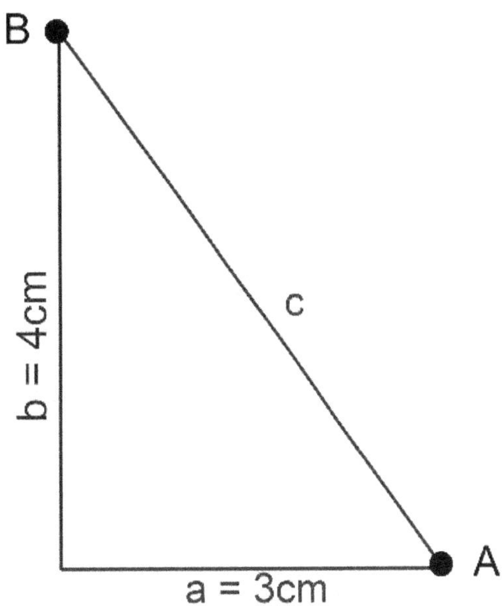

Figure 1.17.

$$\boldsymbol{a^2 + b^2 = c^2}$$

$$\Rightarrow \boldsymbol{c = \sqrt{a^2 + b^2}}$$

Therefore, the solution to our example would be:

$$(3cm)^2 + (4cm)^2 = c^2$$

$$\Rightarrow 25cm^2 = c^2 \Rightarrow c = 5cm$$

What if we know the coordinates of points A and B? Can we then find their distance? Well, it is the same story; we need to know how far we need to go left or right vs. up or down and then use the aforementioned formula to find the distance.

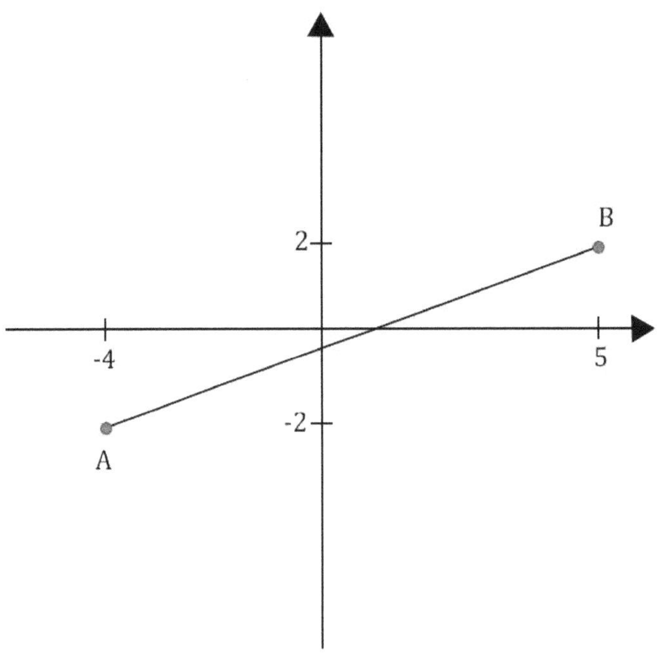

Figure 1.18.

Now the question is, how do we find how far we need to go left or right? Well, let's say I am 28 years old, and you are 16. How much older am I than you? 12? Correct! How did you find it? Good job, subtract my age from your age. Now let's get back to our problem How far did we go left or right from point A? Isn't it the x-coordinate of the point B minus the x-coordinate of the point A? Why x-coordinate? The reason is the x-coordinate is all about left and right movement.

That is, $X_B - X_A$ or $X_{second\ point} - X_{first\ point}$ is the horizontal distance from point A to B. How about the up and down part? Brilliant, the y coordinate is about up and down movement. Therefore, $Y_B - Y_A$or $Y_{second\ point} - Y_{first\ point}$ would give the vertical distance between point A and B.

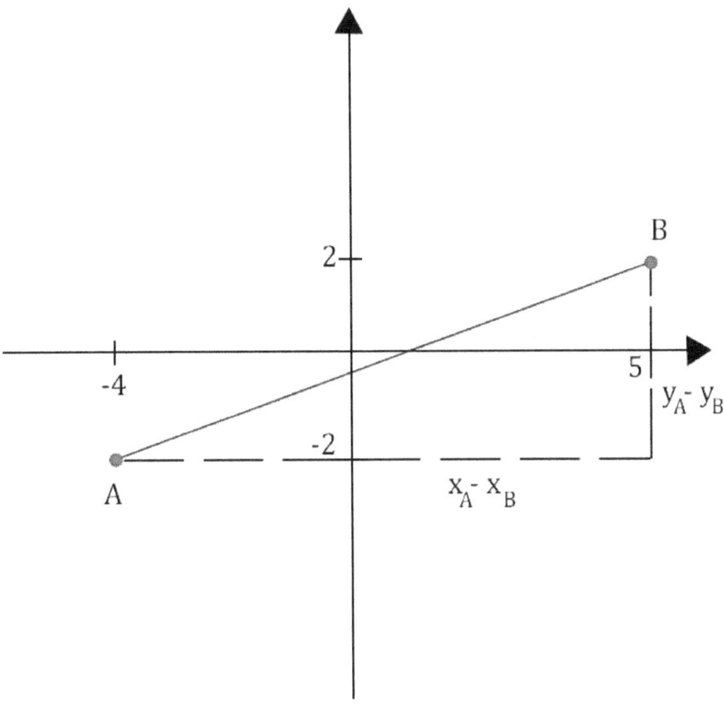

Figure 1.19.

Now let's get back to the Pythagorean formula and use it to find the distance between the two points given our knowledge of the coordinate of the points.

$c = \sqrt{a^2 + b^2}$ given horizontal movement (left or right displacement) equals to $a = X_B - X_A$, and vertical motion is equivalent to (up or down changes) $Y_B - Y_A$. We can adjust the formula as follows:

$$\textbf{Distance between the two points} = \sqrt{(X_B - X_A)^2 + (Y_B - Y_A)^2}$$

Example:

---

15. Find the distance between point A=(-4,-2) and the point B=(5,2) as presented in the previous figure.

$$\sqrt{(X_B - X_A)^2 + (Y_B - Y_A)^2} = \sqrt{\left(5-(-4)\right)^2 + \left(2-(-2)\right)^2} = \sqrt{9^2 + 4^2} = \sqrt{97}$$

F.A.Q. (Frequently Asked Questions)

1. **Do we always need to have at least one right angle to use the Pythagorean Theorem?**

   Yes. However, you will see that we have a more generalized formula to handle all kinds of triangles.

2. **Would the result always be positive for a, b, c?**

   Yes. The result has to be positive because the theorem is about length, which is always positive. Moreover, as we have mentioned previously, the square root of a number is always a positive value, unless we specify otherwise.

3. **Can I turn the triangle and still use the theorem?**

   Yes. All you have to do is keep the longest side in mind (the side facing the 90° or the side that does not have the 90° on its corners). The longest side's length squared is equal to the addition of the magnitude of other sides' squared. Keep in mind that C is always the value of the longest side, which is also known as the hypotenuse.

$$a^2 + b^2 = c^2$$ → C is always the value of the longest side

4. **For the distance formula $\sqrt{(X_B - X_A)^2 + (Y_B - Y_A)^2}$, how do we know which point is A and which is B?**

Regardless of your choice, you would obtain the same result. The square and the square root of real values are positive.

1.3 $\sin^2(x) + \cos^2(x) = 1$

Now try to find the length of r by calculating the distance from point A to point B and from point A to point C.

Solution:

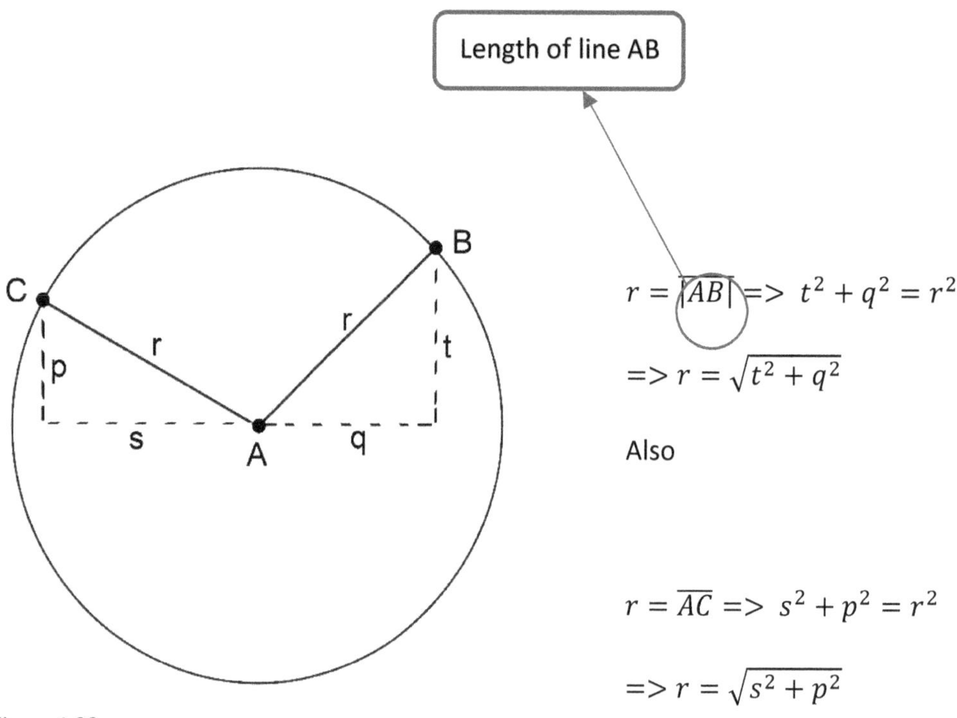

Figure 1.20.

In an ordinary condition, do we agree with the following sentence? If, at a particular time, you are as tall as Alex, and you are as tall as Jake, then Alex is as tall as Jake. Right? Now do you agree that if two lengths are equal to the same thing (in our case r or r^2), they are equal to each other? Therefore, we can conclude:

$$\sqrt{s^2 + p^2} = \sqrt{t^2 + q^2}$$

Or

$$s^2 + p^2 = t^2 + q^2$$

Side Note: It is a common practice to show the points with capital letters while representing the lines with lower case letters. Given the information presented, we know the distance from two points is given by the formula provided. Now think about a circle. Wouldn't you define a circle as a set of points that have equal distance from a central point on a plane? This same length is known to be the radius.

Do you agree that every single point that we choose on the circumference of the circle would have the same distance from the center (equal to r), and they would be equal to each other?

Now let's go back to our sin(x) and cos(x) coordinate.

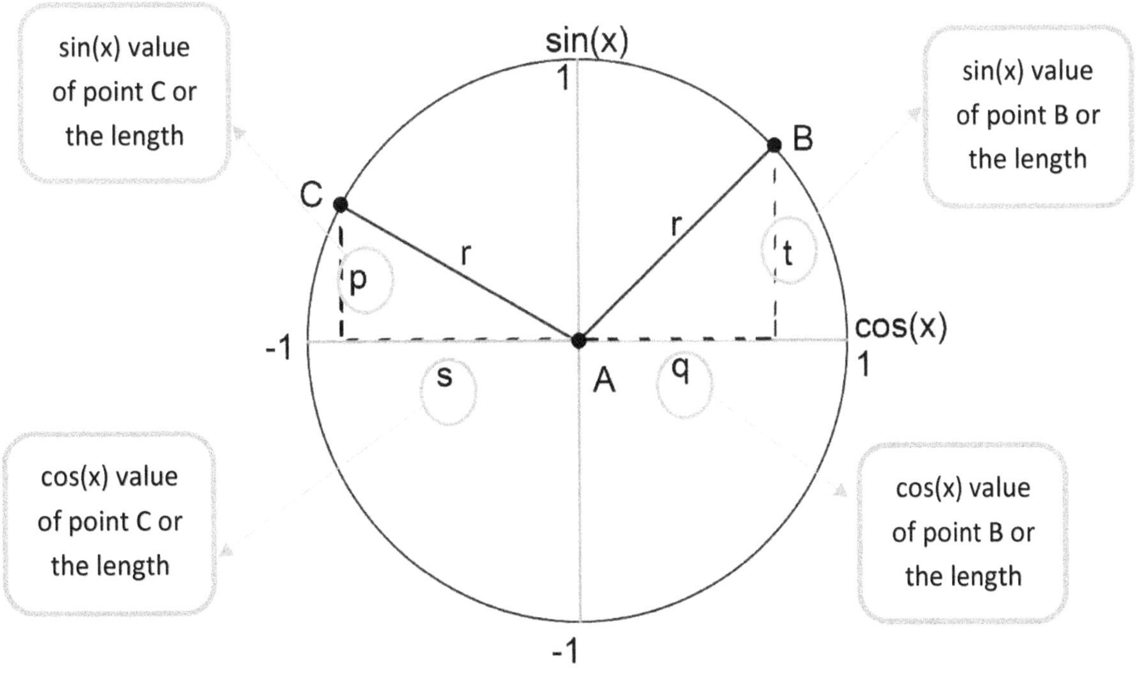

Figure 1.21.

So we would have the following facts for point the B on the circle:

$r = \overline{AB} => t^2 + q^2 = \sin^2(x) + \cos^2(x) = r^2$

We know r is equal to 1, so the aforementioned equation would be equal to 1 as well.

$\sin^2(x) + \cos^2(x) = 1^2 = 1 => \sin^2(x) + \cos^2(x) = 1$

Also, we would have the following equation for point C on the circumference of the circle:

$r = \overline{AC} => s^2 + p^2 = \sin^2(x) + \cos^2(x) = 1$

If we try this at any other points, we get the same result. Since we can only choose points that are on the circumference of the circle, we always get the same equation.

$\sin^2(x) + \cos^2(x) = 1$

Apparently, it is considered to be a big deal! So I want to write the same formula again, but this time with a larger font and with red.

$$\sin^2(x) + \cos^2(x) = 1$$

F.A.Q.

---

1. **Why did we choose to have a circle with a radius of 1? Why did we choose sin(x) and cos(x)?**

   For simplicity of calculation, we could have opted the radius to be 2 and also name everything differently. The result would have been the same, and we could still have found the ratio of the triangle (or in my opinion any three points.)

2. **I have studied sin(x) and cos(x) in context of triangles not circle(s), so why did you define them with a unit circle?**

We would explain them using triangles as well. I wanted you to learn the context that sin(x) and cos(x) are some actual lengths.

3. **I thought sin(x) and cos(x) were angles, so why do you say they are lengths?**

   sin(x), cos(x), tan(x), cot(x), sec(x), and csc(x) are sets of ratios associated to angle x of a right triangle. In this section, we have defined trigonometric ratios as lengths. For example, when we say sin(x), we mean the magnitude of the line that projected on the sin(x) axis while we are x degrees counterclockwise from positive side of cos(x) axes. We will discuss these issues in detail further in the book.

Examples:

Find sin(x) in the following situations:

15. cos(x) = 0.5

Solution:

We use the famous trigonometric identity $\sin(x)^2 + \cos(x)^2 = 1$

So we substitute the value of cos(x) and we get: $\sin(x)^2 + (0.5)^2 = 1$

$$\sin(x)^2 = 1 - (0.5)^2 => \sin(x) = \mp\sqrt{1-(0.5)^2}$$

$$= \mp\sqrt{.75} = \mp\sqrt{\frac{3}{4}} = \mp\frac{\sqrt{3}}{\sqrt{4}} = \mp\frac{\sqrt{3}}{2}$$

Side Note: Keep in mind that the square root of a number is always positive. Please do not confuse that with the two possible roots that result from a degree 2 equation. If $x^2 = 4$ then $x = \mp\sqrt{4} = \mp 2$ but $\sqrt{4} = +2$ only. Confusing? Good!

16. cos(x) = 0

Solution:

We use the famous trigonometric identity $\sin^2(x) + \cos^2(x) = 1$.

So we substitute the value of cos(x) and we get: $\sin^2(x) + (0)^2 = 1$.

$\sin^2(x) = 1 - (0)^2 \Rightarrow \sin(\mathrm{x}) = \mp\sqrt{1-(0)^2} = \mp\sqrt{1} = \mp 1.$

---

17. cos(x) = -1
Solution:

We use the famous trigonometric identity $\sin^2(x) + \cos^2(x) = 1$.

So we substitute value of cos(x) and we get: $\sin^2(x) + (-1)^2 = 1$.

$\sin(\mathrm{x})^2 = 1 - (-1)^2 \Rightarrow \sin(\mathrm{x}) = \mp\sqrt{1-(-1)^2} = \mp\sqrt{0} = \mp 0 = 0.$

---

18. cos(x)= 3

Solution:

Oops! cos(x) would be outside of the circle we have defined. In our circle, cos(x) can only be between -1 and 1, and 3 is not in that range. Don't believe me? Let's solve this and see what happens.

We use the famous trigonometric identity $\sin^2(x) + \cos^2(x) = 1$.
So we substitute value of cos(x) and we get: $\sin^2(x) + (3)^2 = 1$.

$\sin^2(x) = 1 - (0.5)^2 \Rightarrow \sin(\mathrm{x}) = \mp\sqrt{1-(3)^2} = \mp\sqrt{-8}.$

What is the square root of a negative number? It definitely is not defined for real numbers, and we know the sin(x) is between -1 and 1, and from -1 and 1 numbers are real on the sin(x) axis. Therefore, the solution does not exist for sin(x) as we have defined sin(x).

I know you believed me, just wanted to check for myself!

---

19. cos(x)=$\frac{\sqrt{3}}{2}$

Solution:

$$\sin^2(x) + \cos^2(x) = 1$$

$$\sin^2(x) + \left(\frac{\sqrt{3}}{2}\right)^2 = 1$$

$$\sin^2(x) = 1 - \frac{3}{4} = \frac{1}{4}$$

$$\sin(\mathrm{x}) = \pm\sqrt{\frac{1}{4}} = \pm\frac{1}{2}$$

---

20. cos(x)= 1.1

Solution:

The same goes for cos(x), since cos(x) is undefined for any value more than 1, so cos(x) does not exist to produce a corresponding sin(x).

21. cos(x)= -0.5

Solution:

We use the famous trigonometric identity $\sin(\mathrm{x})^2 + \cos(\mathrm{x})^2 = 1.$

So we substitute value of cos(x) and we get: $\sin(\mathrm{x})^2 + (-0.5)^2 = 1.$

$$\sin^2(x) = 1 - (0.5)^2 \Rightarrow \sin(\mathrm{x}) = \mp\sqrt{1 - (-0.5)^2} = \mp\sqrt{.75} = \mp\sqrt{\frac{3}{4}} = \mp\frac{\sqrt{3}}{\sqrt{4}} = \mp\frac{\sqrt{3}}{2}$$

---

22. sin(x)=1

Solution:

Really? If sin(x) =1 then sin(x) =1. What a coincidence!

Not funny!

---

23. cos(x)=0.6

Solution:

$\sin^2(x) + \cos^2(x) = 1$

$\sin^2(x) + (0.6)^2 = 1$

$\sin^2(x) = 1 - (0.6)^2 = 1 - 0.36 = 0.64$

$\sin(\mathrm{x}) = \pm\sqrt{0.64} = \pm 0.8$

1.4 Measuring Angles (Radians and Degrees):

Measuring something requires a unit, for example, the unit of time is seconds. There have been many attempts in which scientists tried to define the unit of time, second, one of them follows. When the atom of Caesium 133's radiation jumps 9,192,631,770, it is equivalent to one second. What about length? It is more complicated since there are different units for it that are widely used. For example, the meter, which is the length that light travels in a vacuum in 1/299,792,458 of a second. On the other side, there is another unit for length called inch, which has the following relationship with the meter:

1 inch = 0.0254 meter

Now my question is, does it mean 1 = 0.054?

Right, of course not! Then why is 1 inch = 0.0254 meter?

The answer is hidden in the unit and it very important. Keep this in mind as we get back to it.

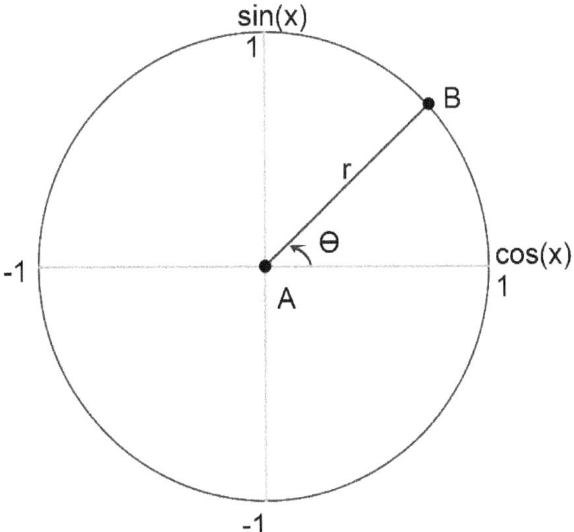

Figure 1.22

Now we move on to measuring angles. But, before that, let's see what I mean by an angle. Imagine you have two lines intercepting at a point. In simple terms, assume an angle is how much you need to turn in order to get to the other line. Before introducing any unit, keep in mind that, if you go counterclockwise, the angle is positive, and negative if the movement is clockwise. If you are ever confused about which side is positive, imagine you are standing outside facing north, and you are looking at the sun as it moves in the sky. It rises in the east (right side), and it sets in the west (left side). This would give you the positive direction.

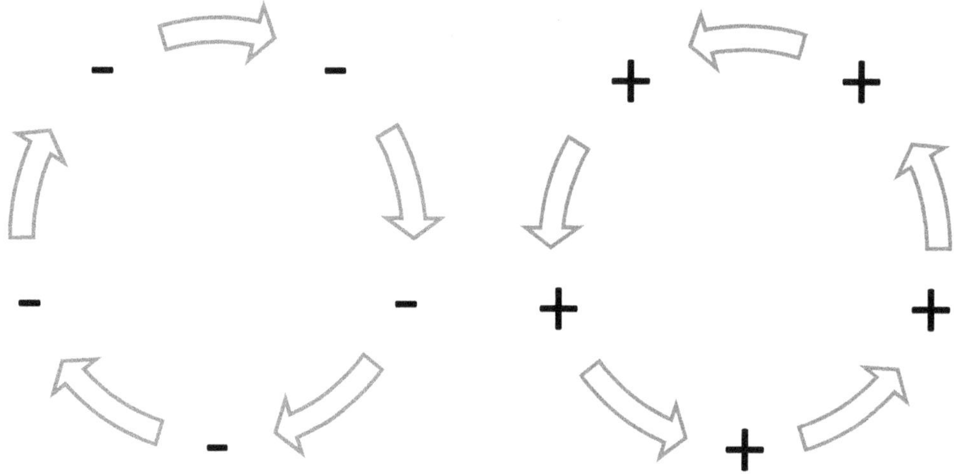

Figure 1.23.

Figure 1.24.

Turn

The most straightforward unit would be a turn, which is a full cycle. There a lot of other units which are defined, such as sextant, hour angle, point, hexacontade, pechus, binary degree, diameter part, grad, mil and the list would go on and on. The basis for all of the units mentioned is the way a turn is divided. In fact, we can come up with our own unit. For example, let's divide a turn into 1,000 equal parts, and we call each of them a "hezarom."

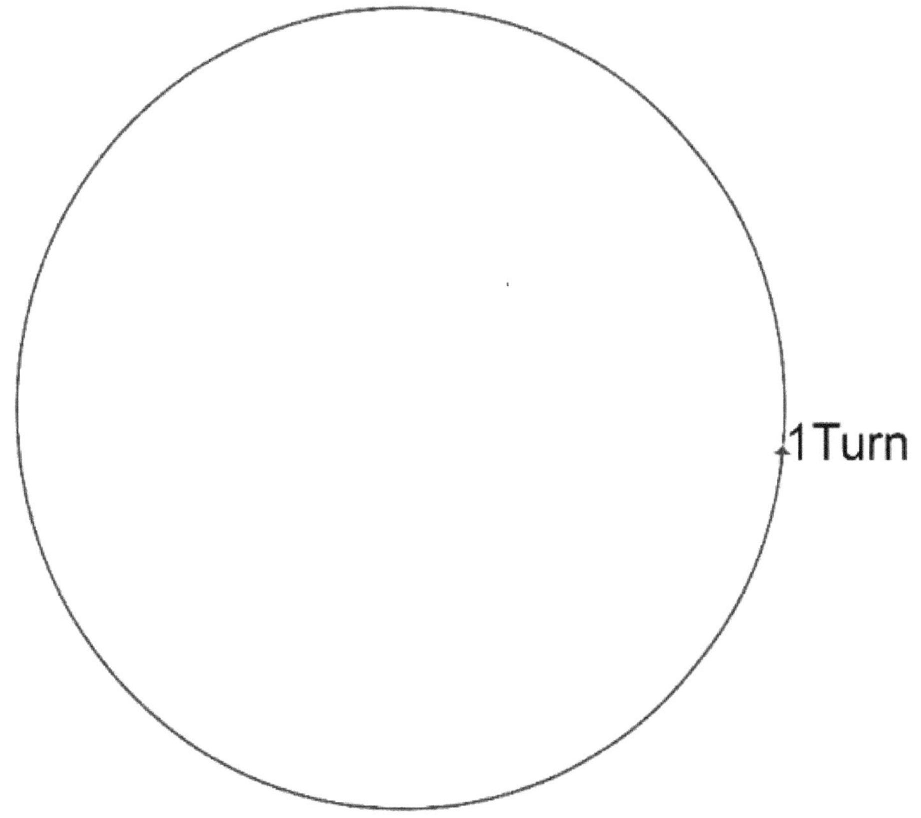

Figure 1.25.

Degree

One of the most widely used units for measuring an angle is the degree. If you divide a turn into 360 equal units, each unit is called a degree. There are a few theories on how it was originated, and one of them is as follows. As you know, the base of our number is 10, and the base for binary numbers (that is widely used in ordinary computers) is 2.

Back in ancient days (3rd millennium BC), Sumerians used a system called Sexagesimal, which means base 60 (to be more precise, base 59). It is actually cool when you look at it. For the Sexagesimal system, they had 59 different symbols as follows. The interesting part is they had no symbol for 0 at the time (later on, they constructed 0).

1	11	21	31	41	51
2	12	22	32	42	52
3	13	23	33	43	53
4	14	24	34	44	54
5	15	25	35	45	55
6	16	26	36	46	56
7	17	27	37	47	57
8	18	28	38	48	58
9	19	29	39	49	59
10	20	30	40	50	

Figure 1.26.

Now, getting back to a turn, the theory explains that Babylonians used equilateral triangles and placed them in a circle. They noticed that they could fit exactly six of them. They further used the Sexagesimal system described before to divide each of the equilateral triangles into 60 sections. As a result, they came up with 6 x 60 equal sections for a total of 360 equal parts.

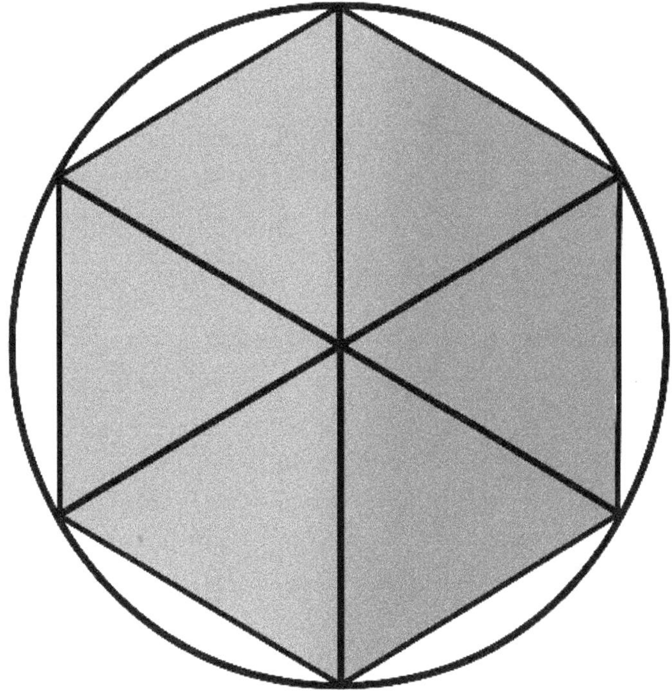

Figure 1.27.

Not to forget how significant of a number 360 is when it comes to divisibility. One of the theories about the origins of 360 degrees is based on the divisibility power of 360. Just a small fact about this incredible number is that it is the smallest number divisible by all numbers from 1 to 10 except for 7.

The other theory for the origin of 360° refers to the approximate number of days in a year and the use of ancient calendars, such as the Persian calendar.

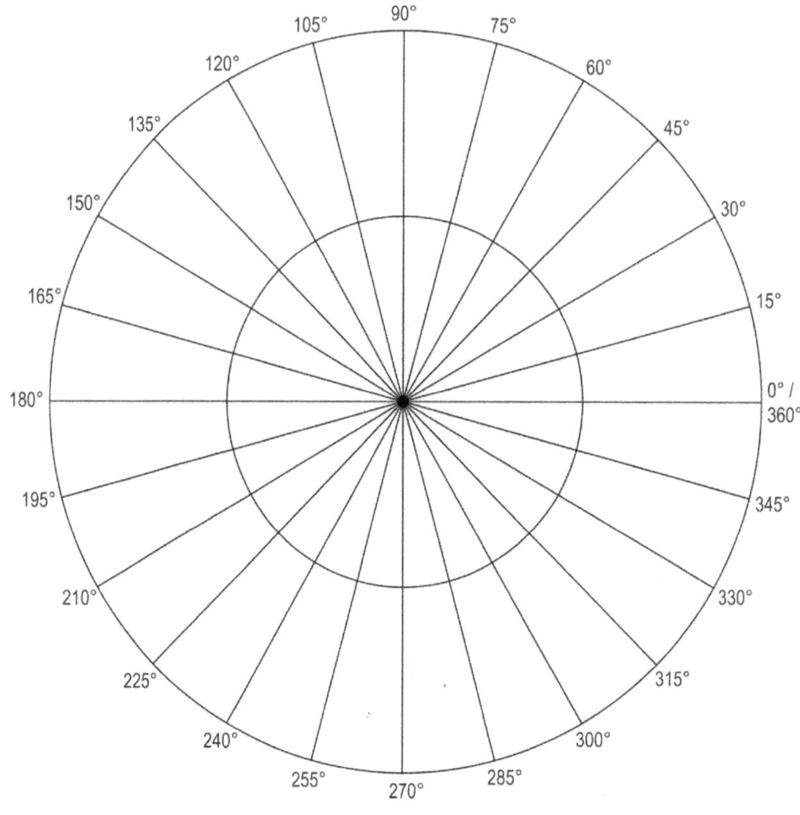

Figure 1.28.

F.A.Q.

1. **Is it possible for an angle to be less than 0 degrees or more than 360 degrees?**

   It is. A circle contains all possible angles. For example, if we try to find 450°, then we need to complete a full turn and add 90°. It is as if we are just showing 90°.

2. **Are angles such as 90° and 450°, which differ by an exact multiplicative of 360° (one or more full turn difference), the same?**

Technically, they are not the same, but, since they are showing the same position, they can be considered equal. Imagine that you're on a flight from your city to the North Pole. If you pass the pole once and make a complete turn and land on the pole the second time, your destination is the same; however, you have traveled more distance.

3. Which one is larger, 1 meter or 30° angle of a circle with radius .5 meters?

2 kilograms! They have different units so you cannot compare them. It is not comparable to say which one is more, two apples or three oranges? One has more apples than the other, and one has more oranges. In other words, you can only compare two quantities if they have identical units.

Examples:

24. How many degrees are in a half turn?

Solution:

We know: 1 turn = 360°, and we also know half is $\frac{1}{2}$ or simply 0.5.

$$\frac{1\ turn}{0.5\ turn} = \frac{360^o}{x} => x = \frac{0.5\ \cancel{turn}\ .\ 360^o}{1\ \cancel{turn}} = 180^o$$

Please note how we eliminated the unit "turn" and ended up with the degrees as the final unit.

25. How many degrees are in $\frac{3}{4}$ turns?

Solution:

We know: 1 turn = 360° and we also know $\frac{3}{4}$ is 0.75.

$$\frac{1\ turn}{0.75\ turn} = \frac{360^o}{x} => x = \frac{0.75\ \cancel{turn}\ .\ 360^o}{1\ \cancel{turn}} = 270^o$$

---

26. How many degrees are in $\frac{1}{4}$ turns give you are going clockwise?

Solution:

$\frac{1}{4}$ turns in a clockwise direction is -0.25 turn (clockwise = negative).

$$\frac{1\ turn}{-0.25\ turn} = \frac{360^o}{x} => x = \frac{-0.25\ \cancel{turn}\ .\ 360^o}{1\ \cancel{turn}} = -90^o$$

Radian

Radian is a beautiful unit that describes an angle in a circle by dividing arc length (in our example length of AB) by the radius.

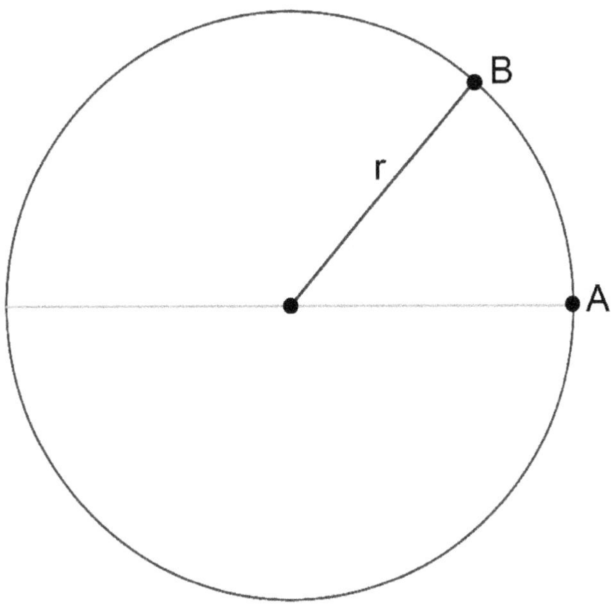

Figure 1.29.

That is, in our example, if length AB equals the radius, then the resulting angle would equal one radian (1 rad).

By Definition:

$$\frac{\text{radius}}{\text{radius}} = 1\,rad$$

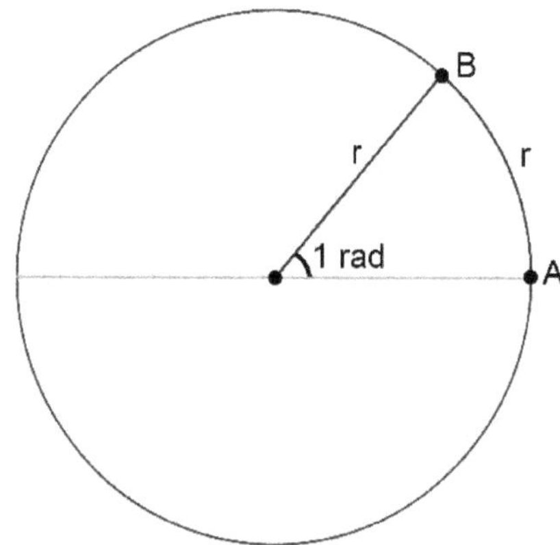

Figure 1.30.

Now that we know what 1 rad is, let's try to find out how many radians a full turn contains. To do that, we need to calculate the number of radii required to cover the circumference of a circle. The following graph illustrates that a half circle would have a number close to 3.14 radii length, which is a bit more than three times the radius' length.

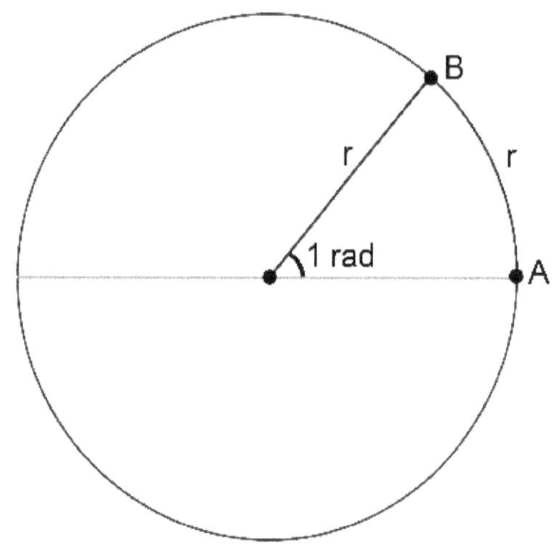

Figure 1.31.

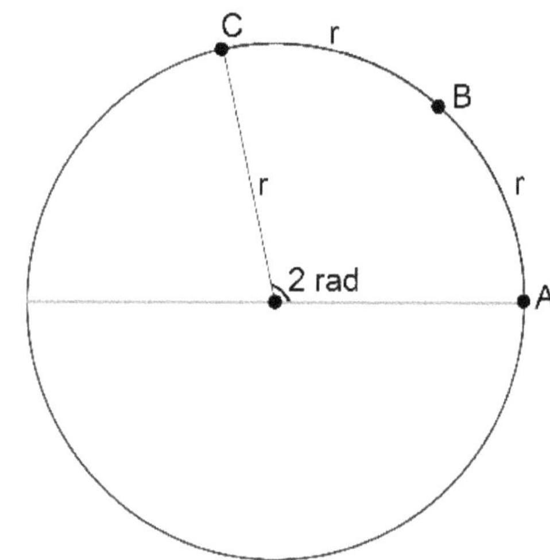

Figure 1.32.

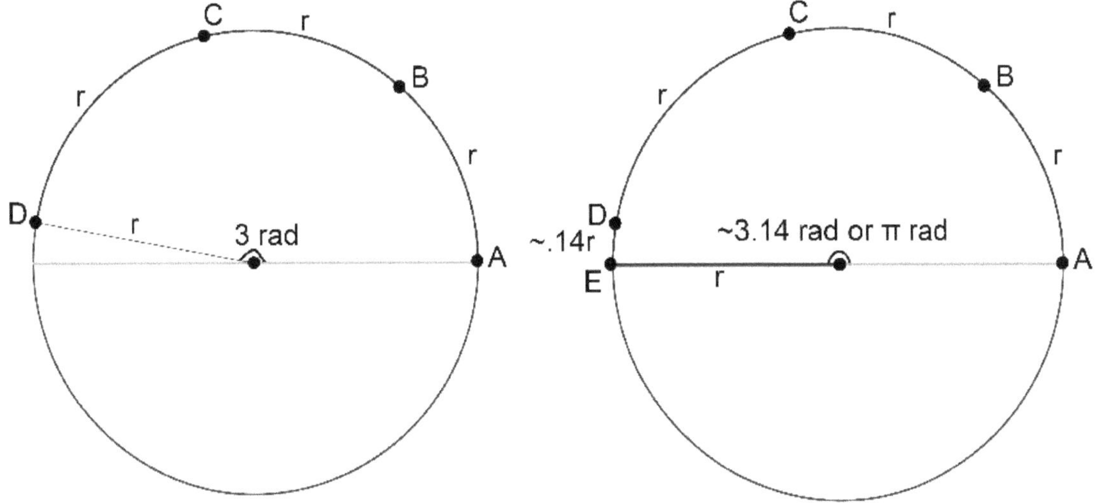

Figure 1.33.

Figure 1.34.

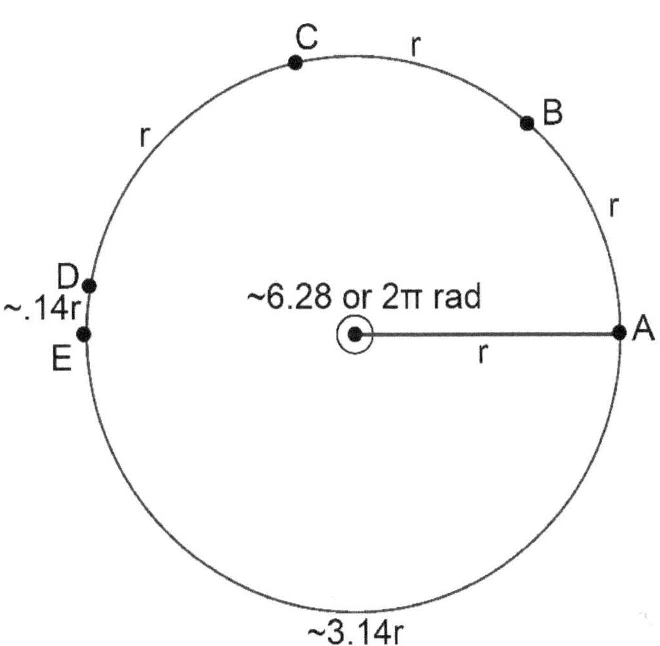

Figure 1.35.

Measuring angles using the radian as opposed to the degrees is a relatively new idea. Roger Cotes defined radian in 1714. However, measuring the angle using arc length was already in used in 1400 AD by mathematicians such as Kashani (Persian mathematician also known as al-Kashi).

Kashani estimated π to 17 decimal places, which was more accurate than the earlier estimation. Interestingly, it took the scientific community close to 200 years to verify his correct estimation, which was done by Ludolph Van Ceulen (20 decimal places).

F.A.Q.

1. **What is one turn equal to in radian units?**

   2π radian is equal to a full turn.

2. **Is 360° equal to 2π rad?**

   Yes.

3. **How do we show units of rad?**

   Unit of radian can be shown as "rad" or even with nothing. Many resources correctly show <u>2π rad</u> as <u>2π</u>.

4. **Which is larger, π rad or 5 rad?**

   5 rad, remember π is just a number approximately equal to 3.14.

5. **We have mentioned in question 2 that 360° equals 2π. Does it mean 2π equals 360 in trigonometry?**

   Absolutely not! 360 is 360 and 2π is almost equal to 3.14. The reason we say that 360° equals 2π is because we do not write down the unit for rad. That is, if we wanted to explain further, we could say 360° equals 2π rad. The two have different units. As we have explained in question 3, it is possible not to show units of radian (rad).

Examples:

Convert the following to Radians:

25. 1 turn

Solution:

1 turn = 2 π

---

26. $\frac{3}{4}$ turn

Solution:

1 turn = 2 π so $\frac{1\ turn}{2\pi} = \frac{\frac{3}{4}turn}{x} => x = \frac{2\pi\ .\frac{3}{4}\cancel{turn}}{1\ \cancel{turn}} = \frac{3}{2}\pi$

---

27. $\frac{7}{8}$ turn

Solution:

1 turn = 2 π so $\frac{1\ turn}{2\pi} = \frac{\frac{7}{8}turn}{x} => x = \frac{2\pi\ .\frac{7}{8}\cancel{turn}}{1\ \cancel{turn}} = \frac{7}{4}\pi$

---

28. 60°

Solution:

360° = 2 π so $\frac{360^o}{2\pi} = \frac{60^o}{x} => x = \frac{2\pi .60^o}{360^o} = \frac{1}{3}\pi = \frac{\pi}{3}$

---

29. 45°

Solution:

360° = 2 π so $\frac{360^o}{2\pi} = \frac{45^o}{x} => x = \frac{2\pi .45^o}{360^o} = \frac{1}{4}\pi = \frac{\pi}{4}$

---

30. 30°.

Solution:

360° = 2 π so $\frac{360^o}{2\pi} = \frac{30^o}{x} => x = \frac{2\pi .30^o}{360^o} = \frac{1}{6}\pi = \frac{\pi}{6}$

---

31. 90°

Solution:

360° = 2 π so $\frac{360^o}{2\pi} = \frac{90^o}{x} => x = \frac{2\pi .90^o}{360^o} = \frac{1}{2}\pi = \frac{\pi}{2}$

---

32. 180°

Solution:

360° = 2 π so $\frac{360^o}{2\pi} = \frac{180^o}{x} => x = \frac{2\pi.180^o}{360^o} = \frac{1}{1}\pi = \pi$

---

33. 270°

Solution:

360° = 2 π so $\frac{360^o}{2\pi} = \frac{270^o}{x} => x = \frac{2\pi.270^o}{360^o} = \frac{3}{2}\pi = \frac{3\pi}{2}$

---

34. 225°

Solution:

360° = 2 π so $\frac{360^o}{2\pi} = \frac{225^o}{x} => x = \frac{2\pi.225^o}{360^o} = \frac{5}{4}\pi = \frac{5\pi}{4}$

---

35. 315°

Solution:

$360^o = 2\pi$ so $\frac{360^o}{2\pi} = \frac{315^o}{x} \Rightarrow x = \frac{2\pi .315^o}{360^o} = \frac{7}{4}\pi = \frac{7\pi}{4}$

---

36. -135°

Solution:

$360^o = 2\pi$ so $\frac{360^o}{2\pi} = \frac{-135^o}{x} \Rightarrow x = \frac{2\pi .(-135^o)}{360^o} = \frac{-3}{4}\pi = \frac{-3\pi}{4}$

---

37. -225°

Solution:

$360^o = 2\pi$ so $\frac{360^o}{2\pi} = \frac{-225^o}{x} \Rightarrow x = \frac{2\pi .(-225^o)}{360^o} = \frac{-5}{4}\pi = \frac{-5\pi}{4}$

---

38. -450°

Solution:

$360^o = 2\pi$ so $\frac{360^o}{2\pi} = \frac{-450^o}{x} \Rightarrow x = \frac{2\pi .(-450^o)}{360^o} = \frac{-5}{2}\pi = \frac{-5\pi}{2}$

Convert the following angles from radians to degrees (please and thank you):

39. 2π

Solution:

$2\ \pi = 360^\circ$

---

40. $-\frac{\pi}{4}$

Solution:

$360^\circ = 2\ \pi \Rightarrow \pi = 180^\circ$. An easy method would be, anywhere you see π, replace it with 180°

$\Rightarrow -\frac{180^O}{4} = -45^\circ$

> **Side Note:** please be advised, even though we say π, we mean π rad. Otherwise, π by itself is approximately 3.14.

---

41. $\frac{\pi}{3}$

Solution:

$180^\circ = \pi \Rightarrow \frac{180^O}{3} = 60^O$

---

42. $\frac{3\pi}{4}$

Solution:

$180^{\circ} = \pi => \frac{3\times180^{o}}{4} = 135^{o}$

---

43. $\frac{\pi}{6}$

Solution:

$180^{\circ} = \pi => \frac{180^{o}}{6} = 30^{o}$

---

44. $-\frac{\pi}{9}$

Solution:

$180^{\circ} = \pi => -\frac{180^{o}}{9} = -20^{o}$

---

45. $\frac{5\pi}{18}$

Solution:

$180^{\circ} = \pi => \frac{5\times180^{o}}{18} = 50^{o}$

Please convert the following angles to an equivalent angle between 0 and 2π or 0° to 360°.

46. $\frac{18\pi}{5}$

Solution: The idea is we need to remove as much 2π or $360°$ as possible from the original angle until we get the smallest positive number. One easy method would be to follow the following steps:

1. Divide the angle by 2π (if in radian) or by $360°$ (if in degree). Disregard the remainder and only consider the quotient part, which is a whole number.

$$\frac{\frac{18\pi}{5}}{2\pi} = \frac{18\pi}{5} \times \frac{1}{2\pi} = \frac{9}{5} = 1 + \frac{4}{5}$$

Note that we only need the quotient part, which, in our example, is one.

Side Note: You always want the remainder to be the smallest positive number.

2. Subtract $2\pi \times$ the whole number quotient (in our example 1) from the original angle:

$$\frac{18\pi}{5} - (2\pi \times 1) = \frac{18\pi}{5} - (2\pi) = \frac{18\pi}{5} - \frac{10\pi}{5} = \frac{8\pi}{5}$$

The resulting angle is between 0 and 2π.

---

47. 3π

Solution:

$$\frac{3\pi}{2\pi} = \frac{3}{2} = 1 + \frac{1}{2} \Rightarrow 3\pi - (2\pi \times 1) = 3\pi - 2\pi = \pi$$

---

48. $\frac{9\pi}{4}$

Solution:

$$\frac{\frac{9\pi}{4}}{2\pi} = \frac{9\pi}{4} \times \frac{1}{2\pi} = \frac{9}{8} = 1 + \frac{1}{8}$$

$$\frac{9\pi}{4} - (2\pi \times 1) = \frac{9\pi}{4} - (2\pi) = \frac{9\pi}{4} - \frac{8\pi}{4} = \frac{\pi}{4}$$

---

49. $\frac{17\pi}{2}$

Solution:

$$\frac{\frac{17\pi}{2}}{2\pi} = \frac{17\pi}{2} \times \frac{1}{2\pi} = \frac{17}{4} = 4 + \frac{1}{4}$$

$$\frac{17\pi}{2} - (2\pi \times 4) = \frac{17\pi}{2} - (8\pi) = \frac{17\pi}{2} - \frac{16\pi}{2} = \frac{\pi}{2}$$

---

50. $\frac{7\pi}{3}$

Solution:

$$\frac{\frac{7\pi}{3}}{2\pi} = \frac{7\pi}{3} \times \frac{1}{2\pi} = \frac{7}{6} = 1 + \frac{1}{6}$$

$$\frac{7\pi}{3} - (2\pi \times 1) = \frac{7\pi}{3} - (2\pi) = \frac{7\pi}{3} - \frac{6\pi}{3} = \frac{\pi}{3}$$

---

51. -2π

Solution:

$$\frac{-2\pi}{2\pi} = \frac{-1}{1} = -1 + 0$$

$$-2\pi - \left(2\pi \times (-1)\right) = -2\pi + (2\pi) = 0$$

---

4. $\frac{-3\pi}{4}$

Solution:

$$\frac{\frac{-3\pi}{4}}{2\pi} = \frac{-3\pi}{4} \times \frac{1}{2\pi} = \frac{-3}{8} = -1 + \frac{5}{8}$$

Note that the remainder $\left(\frac{5}{8}\right)$ is a positive number, and you cannot accept negative remainders for this method.

$$\frac{-3\pi}{4} - \left(2\pi \times (-1)\right) = \frac{-3\pi}{4} + (2\pi) = \frac{-3\pi}{4} - \frac{8\pi}{4} = \frac{5\pi}{4}$$

---

52. 480°

Solution:

$$\frac{480^o}{360^o} = \frac{4}{3} = 1 + \frac{1}{3}$$

$$480^o - \left(360^o \times (1)\right) = 480^o - (360^o) = 120^o$$

---

53. 4 π°

Solution:

$$\frac{4\pi^{\,o}}{360^o} = \frac{4\pi}{90} = 0 + \frac{4\pi}{90}$$

Remember π is just a number approximately equal to 3.14. Accordingly, 4 π is close to 12.52

$$4\,\pi^o - \left(360^o \times (0)\right) = 4\,\pi^o$$

---

54. 3645°

Solution:

$$\frac{3645^o}{360^o} = \frac{81}{8} = 10 + \frac{1}{8}$$

$$3645^o - \left(360^o \times (10)\right) = 3645^o - (3600^o) = 45^o$$

---

55. -120°

Solution:

$$\frac{-120^o}{360^o} = \frac{-1}{3} = -1 + \frac{2}{3}$$

$$-120^o - (360^o \times (-1)) = -120^o + (360^o) = 240^o$$

---

56. 540°

Solution:

$$\frac{540^o}{360^o} = \frac{3}{2} = 1 + \frac{1}{2}$$

$$540^o - (360^o \times (1)) = 540^o - (360^o) = 180^o$$

---

57. 150π°

Solution:

π is greater than 3 and less than 4, so 5π is more than 15 and less than 20.

$$\frac{150\pi^o}{360^o} = \frac{5\pi}{12} => 1 + \text{something positive and less than } 1$$

$$150\pi^o - (360^o \times (1)) = 150\pi^o - (360^o) = \text{a number close to } 111.24^o$$

1.5 How to Remember Famous Angles

How to remember sin(x) and cos(x) of famous angles:

Step One: Prepare a quick table as follows:

Angle in Degrees	0	30	45	60	90
Angle in Radians	0	$\frac{\pi}{6}$	$\frac{\pi}{4}$	$\frac{\pi}{3}$	$\frac{\pi}{2}$
sin(x)					
cos(x)					

Table 1.1.

Step two:

Angle in Degrees	0	30	45	60	90
Angle in Radians	0	$\frac{\pi}{6}$	$\frac{\pi}{4}$	$\frac{\pi}{3}$	$\frac{\pi}{2}$
sin(x)	$\frac{\sqrt{\ }}{2}$	$\frac{\sqrt{\ }}{2}$	$\frac{\sqrt{\ }}{2}$	$\frac{\sqrt{\ }}{2}$	$\frac{\sqrt{\ }}{2}$
cos(x)	$\frac{\sqrt{\ }}{2}$	$\frac{\sqrt{\ }}{2}$	$\frac{\sqrt{\ }}{2}$	$\frac{\sqrt{\ }}{2}$	$\frac{\sqrt{\ }}{2}$

Table 1.2.

Step 3: From the unit circle, we know sin (0) = 0. As a result, we place the number from 0 to 4, and we repeat for cos (x) backwards as if we are going around the loop.

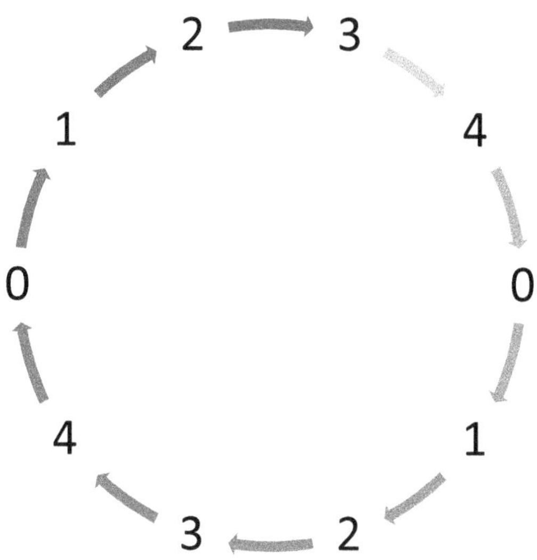

Figure 1.36.

So we get the following numbers:

Angle in Degrees	0	30	45	60	90
Angle in Radians	0	$\frac{\pi}{6}$	$\frac{\pi}{4}$	$\frac{\pi}{3}$	$\frac{\pi}{2}$
sin(x)	$\frac{\sqrt{0}}{2}$	$\frac{\sqrt{1}}{2}$	$\frac{\sqrt{2}}{2}$	$\frac{\sqrt{3}}{2}$	$\frac{\sqrt{4}}{2}$
cos(x)	$\frac{\sqrt{4}}{2}$	$\frac{\sqrt{3}}{2}$	$\frac{\sqrt{2}}{2}$	$\frac{\sqrt{1}}{2}$	$\frac{\sqrt{0}}{2}$

Table 1.3.

As you may have noticed the following are true:

$$\frac{\sqrt{0}}{2} = \frac{0}{2} = 0\,, \frac{\sqrt{1}}{2} = \frac{1}{2}\,, \frac{\sqrt{4}}{2} = \frac{2}{2} = 1$$

Side note: You might be asking if the square root has two answers. For example, would $\sqrt{4} = \pm 2$. We only accept the positive answer. Please refer to the following graph to clarify:

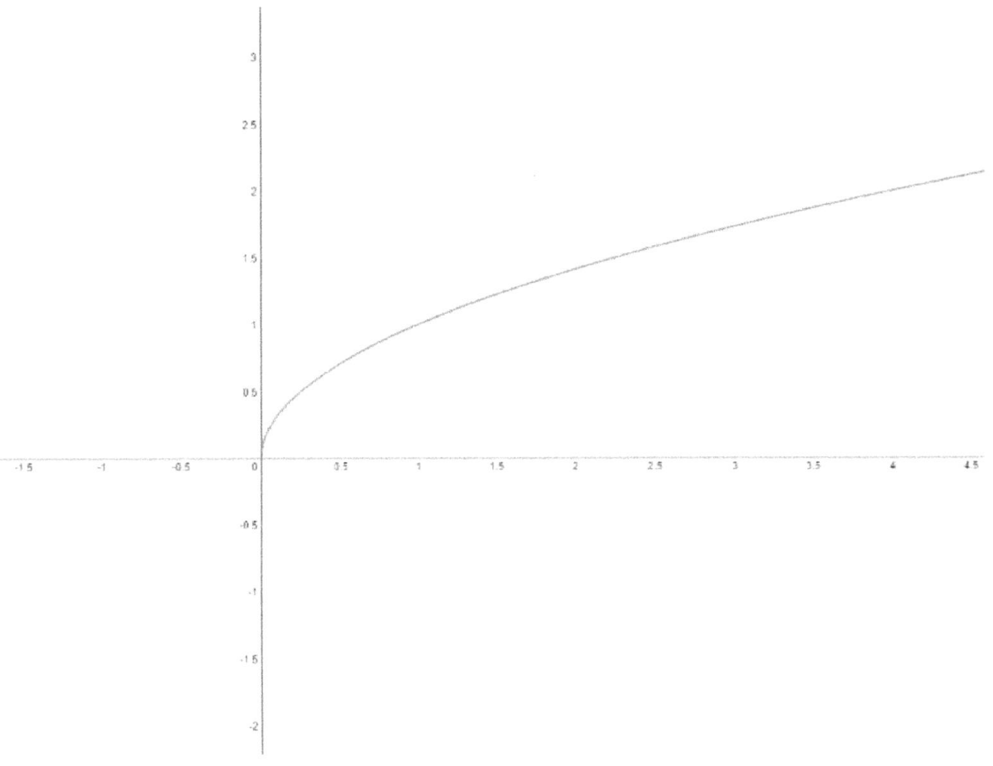

Figure 1.37.

Alternatively, we can say the graph of $y = \sqrt{x}$ does not contain the negative results (the lower side does not exist).

So the table can be rewritten as:

Angle in Degrees	0	30	45	60	90
Angle in Radians	0	$\frac{\pi}{6}$	$\frac{\pi}{4}$	$\frac{\pi}{3}$	$\frac{\pi}{2}$
sin(x)	0	$\frac{1}{2}$	$\frac{\sqrt{2}}{2}$	$\frac{\sqrt{3}}{2}$	1
cos(x)	1	$\frac{\sqrt{3}}{2}$	$\frac{\sqrt{2}}{2}$	$\frac{1}{2}$	0

Table 1.4.

Knowing the value of sin(x) and cos(x), it is easy to extend the table to determine tan(x) and cot(x). Let's first construct tan(x) by dividing the value of sin(x) by cos(x). First, try to insert the values for sin(x) and cos(x) by heart. After that, construct the values for tan(x) by dividing the value of sin(x) by cos(x) (watch what happens if cos(x)=0). You can check your result against the following table:

Angle in Degrees	0	30	45	60	90
Angle in Radians	0	$\frac{\pi}{6}$	$\frac{\pi}{4}$	$\frac{\pi}{3}$	$\frac{\pi}{2}$
sin(x)					
cos(x)					
tan(x)					

Table 1.5.

Please check if the following results are correct and match yours:

Angle in Degrees	0	30	45	60	90
Angle in Radians	0	$\frac{\pi}{6}$	$\frac{\pi}{4}$	$\frac{\pi}{3}$	$\frac{\pi}{2}$
sin(x)	0	$\frac{1}{2}$	$\frac{\sqrt{2}}{2}$	$\frac{\sqrt{3}}{2}$	1
cos(x)	1	$\frac{\sqrt{3}}{2}$	$\frac{\sqrt{2}}{2}$	$\frac{1}{2}$	0
tan(x)	$\frac{0}{1} = 0$	$\frac{\frac{1}{2}}{\frac{\sqrt{3}}{2}} = \frac{1}{\sqrt{3}}$ $= \frac{\sqrt{3}}{3}$	$\frac{\frac{\sqrt{2}}{2}}{\frac{\sqrt{2}}{2}} = 1$	$\frac{\frac{\sqrt{3}}{2}}{\frac{1}{2}} = \sqrt{3}$	$\frac{1}{0} =$ Undefined

Table 1.6.

We can further extend our tables by adding cot(x), csc(x), and sec(x). To find cot(x), you can flip all tan(x) values or divide cos(x) by sin(x). In order to find csc(x), you can flip sin(x) using the definition of csc(x).

Lastly, to find sec(x), you can flip the associated cos(x) value. Using the given information, please fill in the following table and check you answer with mine:

Angle in Degrees	0	30	45	60	90
Angle in Radians	0	$\frac{\pi}{6}$	$\frac{\pi}{4}$	$\frac{\pi}{3}$	$\frac{\pi}{2}$
sin(x)					
cos(x)					
tan(x)					
cot(x)					
csc(x)					
sec(x)					

Table 1.7.

Again, please see if we are correct by checking the following table:

Angle in Degrees	0	30	45	60	90
Angle in Radians	0	$\frac{\pi}{6}$	$\frac{\pi}{4}$	$\frac{\pi}{3}$	$\frac{\pi}{2}$
sin(x)	0	$\frac{1}{2}$	$\frac{\sqrt{2}}{2}$	$\frac{\sqrt{3}}{2}$	1
cos(x)	1	$\frac{\sqrt{3}}{2}$	$\frac{\sqrt{2}}{2}$	$\frac{1}{2}$	0
tan(x)	$\frac{0}{1} = 0$	$\frac{\frac{1}{2}}{\frac{\sqrt{3}}{2}} = \frac{1}{\sqrt{3}}$ $= \frac{\sqrt{3}}{3}$	$\frac{\frac{\sqrt{2}}{2}}{\frac{\sqrt{2}}{2}} = 1$	$\frac{\frac{\sqrt{3}}{2}}{\frac{1}{2}} = \sqrt{3}$	$\frac{1}{0} =$ Undefined
cot(x)	$\frac{1}{0} =$ Undefined	$\frac{\frac{\sqrt{3}}{2}}{\frac{1}{2}} = \sqrt{3}$	$\frac{\frac{\sqrt{2}}{2}}{\frac{\sqrt{2}}{2}} = 1$	$\frac{\frac{1}{2}}{\frac{\sqrt{3}}{2}} = \frac{1}{\sqrt{3}} = \frac{\sqrt{3}}{3}$	$\frac{0}{1} = 0$
csc(x)	$\frac{1}{0} =$ Undefined	$\frac{1}{\frac{1}{2}} = \frac{2}{1} = 2$	$\frac{1}{\frac{\sqrt{2}}{2}} = \frac{2}{\sqrt{2}}$ $= \sqrt{2}$	$\frac{1}{\frac{\sqrt{3}}{2}} = \frac{2}{\sqrt{3}} = \frac{2\sqrt{3}}{3}$	$\frac{1}{1}$=1
sec(x)	$\frac{1}{1}$=1	$\frac{1}{\frac{\sqrt{3}}{2}} = \frac{2}{\sqrt{3}}$ $= \frac{2\sqrt{3}}{3}$	$\frac{1}{\frac{\sqrt{2}}{2}} = \frac{2}{\sqrt{2}}$ $= \sqrt{2}$	$\frac{1}{\frac{1}{2}} = \frac{2}{1} = 2$	$\frac{1}{0} =$ Undefined

Table 1.8.

As you can see, we could construct the entire table given that we know the sin (0). I recommend that you only draw the sin(x) and cos(x) and calculate the rest as you need them.

Examples

Please consider solving the following examples.

59. $\sin(\frac{\pi}{3})\cos(\frac{\pi}{3})$=

Solution:

$$\frac{\sqrt{2}}{2}\times\frac{1}{2}=\frac{\sqrt{2}}{4}$$

---

60. $\sin(\frac{\pi}{2})\cos(0)\tan(\frac{\pi}{2})$=

Solution:

$1\times 1\times 1=1$

---

61. $\sin(\frac{\pi}{6})\cos(\frac{\pi}{2})\cot(\frac{\pi}{2})=\frac{1}{2}\times 0\times 1=0$

Solution:

$\frac{1}{2}\times 0\times 1=0$

---

62. sec(0)+ cos(0)=

Solution:

$1 + 1 = 2$

---

63. csc(sin(0))=

Solution:

$\csc(0) = \frac{1}{0} =$ Undefined

---

64. $\frac{\tan\left(\frac{\pi}{4}\right)\cot^2\left(\frac{\pi}{6}\right)}{\sin^4\left(\frac{\pi}{6}\right)} =$

Solution:

$\frac{1\times(\sqrt{3})^2}{(\frac{1}{2})^4} = \frac{3}{\frac{1}{16}} = 48$

1.6 Constructing Other Angles from Famous Angles

Sin(x) similarities:

We know that sin(x) and cos(x) give us a length. Right? As you noticed, a circle is a very symmetric shape, and sin(x) and cos(x) both have things to do with a circle! So it makes sense that sin(x) and cos(x) each should resemble some symmetric phenomena. So far we know the value of sin(x) and cos(x) given x is between 0° and 90°.

Let's discover some similarities. What is sin(θ) ($0 \leq \theta \leq 90^o$) in the following graph?

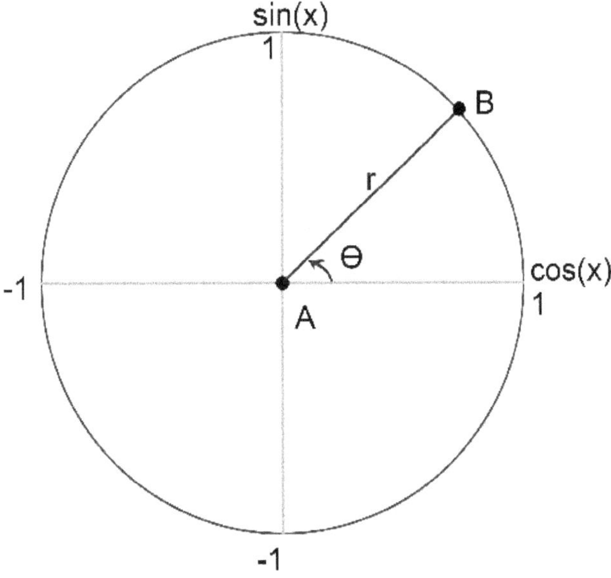

Figure 1.38.

Solution:

As before, we draw a perpendicular line (normal line) to sin(x) axis, and the resulting point shows the magnitude of sin(x). In our example, it would be the length of line AD.

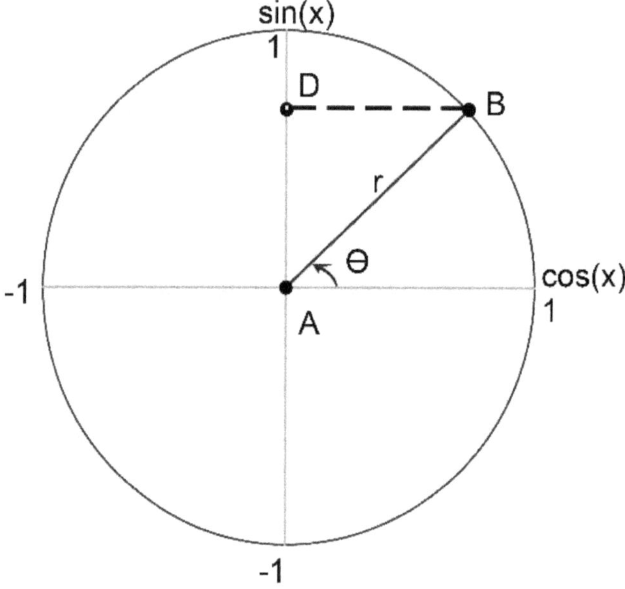

Figure 1.39.

How about the $\sin(\pi - \theta)$?

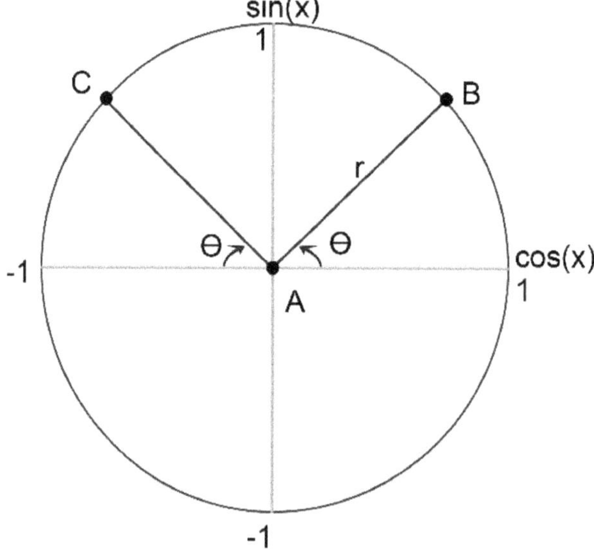

Table 1.40.

Solution:

In the previous section, we draw a normal to sin(x) axis, and we notice we hit point D, the same point we hit last time. Based on the aforementioned logic, we realize sin(π-θ) is also the length of line AD. According to the given relationships, we can intuitively understand the following:

Given θ is in radian:

$$\sin(\pi - \theta) = \sin(\theta)$$

Or if θ is in degree:

$$\sin(180^o - \theta) = \sin(\theta)$$

As an example,

$$\sin(120^o) = \sin(60^o)$$

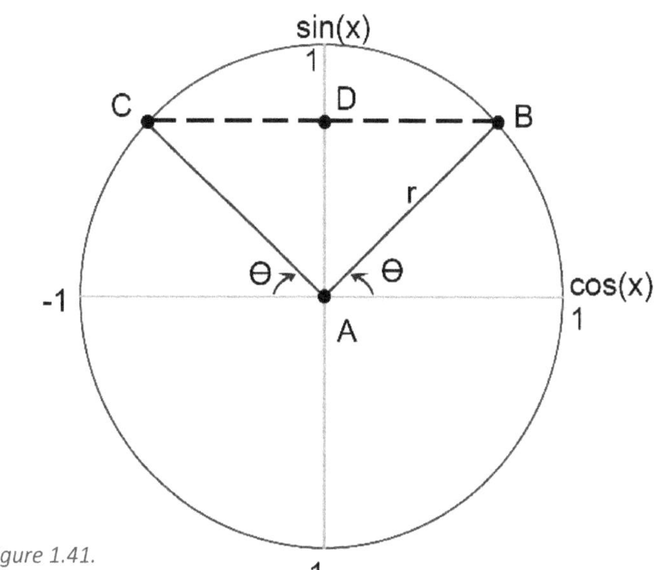

Figure 1.41.

What about $-\theta$, can you find $sin(-\theta)$ in the following example?

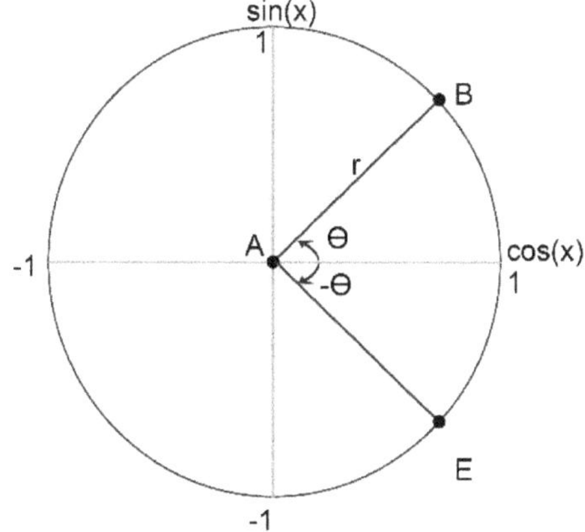

Figure 1.42.

Solution:

To locate the $sin(-\theta)$, we need to draw a normal line to sin(x) axis. In doing so, we intersect the sin(x) axis at the point F. The length of line AF would be equal to the magnitude of $sin(-\theta)$.

Since the value of $sin(-\theta)$ lies in the negative part of the sin(x) axis, then we know the value of $sin(-\theta)$ is the length of line AF. We can see the length of lines AF and AD are equal.

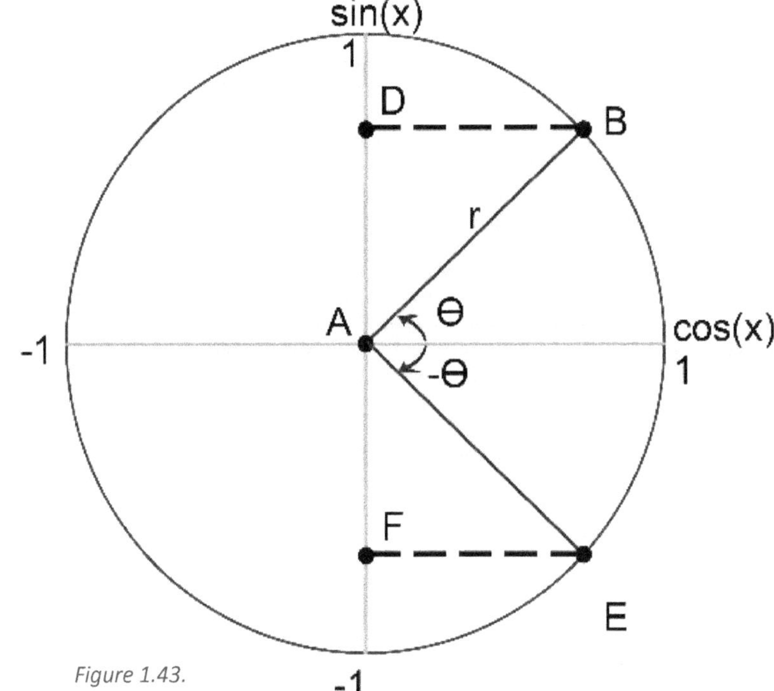

Figure 1.43.

Why? The two triangles Δ AFE and Δ ADB are equal since they each have an equal side (AE and AB are both the radius of the circle) and two equal angles. $\widehat{ADB}$ equals $\widehat{AFE}$ since they both result from perpendicular lines DB and EF, respectively, and they equal 90°. Also, $\widehat{DAB}$ equals $\widehat{FAE}$ as they are both equal to 90°−θ (because the two axes are perpendicular to one another). As a result, we can conclude that the length of AF is equal to the length of AD.

$$sin(\theta) = length\ AD$$

$$sin(-\theta) = -length\ of\ AD$$

$$sin(-\theta) = -sin(\theta)$$

Therefore, we can claim $sin(-30^o) = -sin(30^o) = -\frac{1}{2}$

Please note the angle θ we have chosen for both sine and cosine is between 0 and $\frac{\pi}{2}$ (90°). Also, be courteous how we solve the problem if the angle is in the third quadrant (for example sin (240 °).

Examples:

Find the value of the sine for the following angles based on the famous angles.

65. $\sin(150^o)$

Solution:

$$\sin(150^o) = \sin(180^o - 150^o) = \sin(30^o) = \frac{1}{2}$$

---

66. $\sin(300^o)$

Solution:

Adding or subtracting 360° or 2π to the sine or cosine of an angle would not change the result as we get to the same point (period of both sine and cosine of an angle is 360° or 2π.

$\sin(300^o) = \sin(300^o - 360^o) = \sin(-60^o) = -\sin(60^o) = -\frac{\sqrt{3}}{2}$

---

67. $\sin(330^o)$

Solution:

$\sin(330^o) = \sin(330^o - 360^o) = \sin(-30^o) = -\sin(30^o) = -\frac{1}{2}$

---

68. $\sin(240^o)$

Solution:

$\sin(240^o) = \sin(240^o - 360^o) = \sin(-120^o)$

$= -\sin(120^o) = -\sin(180^o - 120^o) = \sin(60^o) = -\frac{\sqrt{3}}{2}$

---

69. $\sin(225^o)$

Solution:

$$\sin(225^o) = \sin(180^o - 225^o) = \sin(-45^o) = -\sin(45^o) = -\frac{\sqrt{2}}{2}$$

---

70. $\sin(\frac{4\pi}{3})$

Solution:

$$\sin\left(\frac{4\pi}{3}\right) = \sin\left(\pi - \frac{4\pi}{3}\right) = \sin\left(\frac{3\pi - 4\pi}{3}\right) = \sin\left(-\frac{\pi}{3}\right) = -\sin\left(\frac{\pi}{3}\right)$$

$$= -\frac{\sqrt{3}}{2}$$

---

71. $\sin(-\frac{3\pi}{4})$

Solution:

$$\sin\left(\frac{-3\pi}{4}\right) = -\sin\left(\frac{3\pi}{4}\right) = -\sin\left(\pi - \left(\frac{3\pi}{4}\right)\right) = -\sin\left(\frac{4\pi - 3\pi}{4}\right)$$

$$= -\sin\left(\frac{\pi}{4}\right) = -\frac{\sqrt{2}}{2}$$

Cos(x) similarities:

Moving along with cos(θ), What is cos(θ) in the following graph?

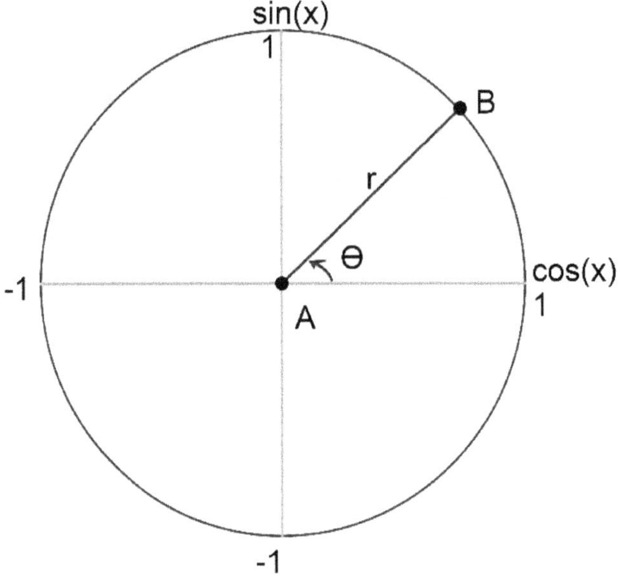

Figure 1.44.

Solution:

Similar to sin(θ), to find cos(θ) we need to draw a perpendicular line to the cos(x) axis. The resulting intersection between the normal line and cos(x) produces a point that we call K. The length of line AK is equal to the cos(θ).

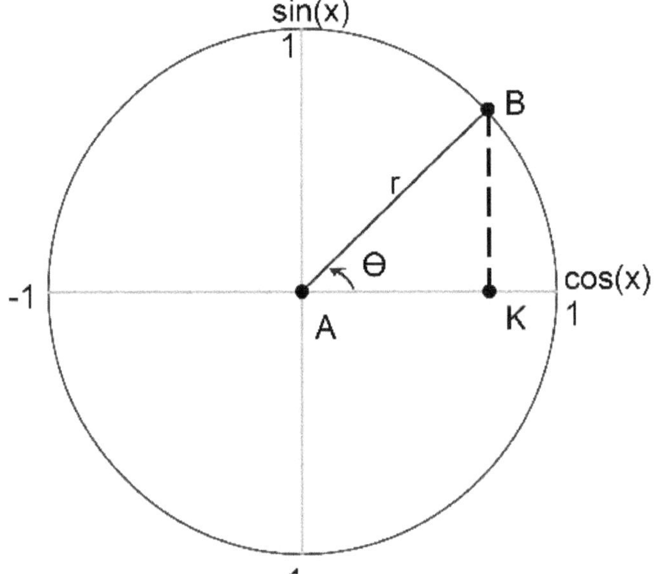

Figure 1.45.

Now, what is cos (-θ) in the following graph?

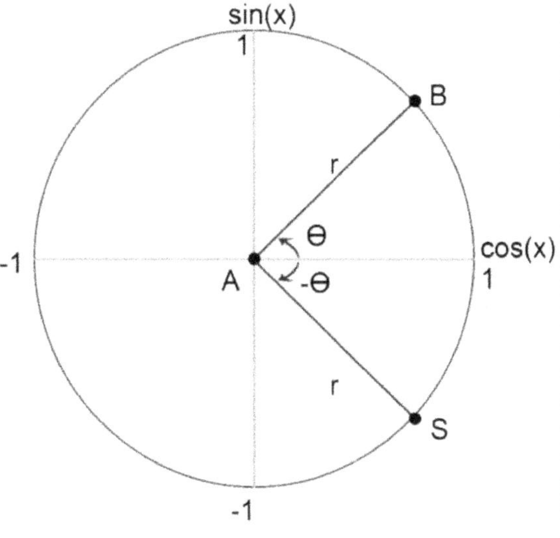

Figure 1.46.

Solution:

Draw a perpendicular line from point S to cos(x) axis and name the resulting intersection K. The length of AK is equal to cos (-θ), which is the same as cos(θ).

As a result, we can conclude:

$$\cos(\theta) = \cos(-\theta)$$

Now let's look at the other side. What is cos $(\pi - \theta)$ it according to the following graph?

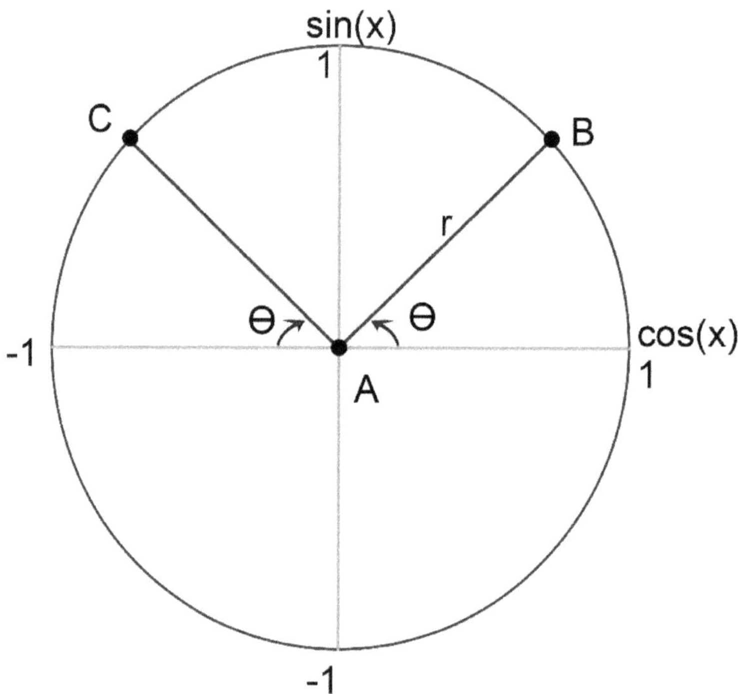

Figure 1.47.

Solution:

To find the $cos(\pi - \theta)$, we need to draw a normal line to the cos (x) axis. We call the intersect point to be T. Length of line AT would be equal to the magnitude of $cos(\pi - \theta)$.

Since the value of $cos(\pi - \theta)$ lies in the negative part of the cos(x) axis, then we know the value of $cos(\pi - \theta)$ is a negative value. Because the length of AT equals the length of AK, we can conclude the following.

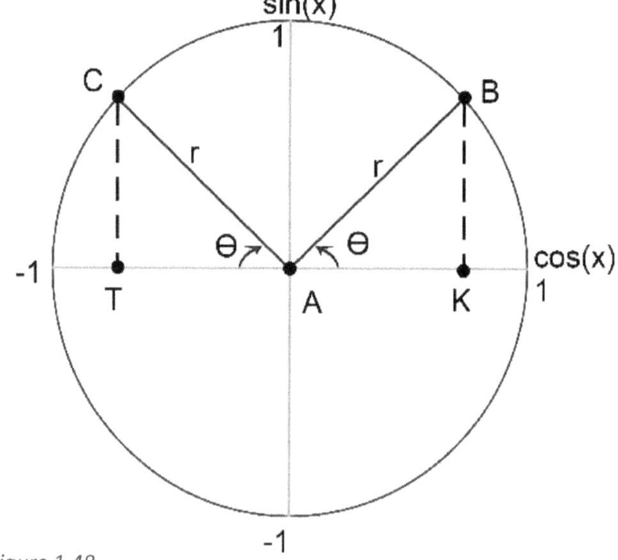

Figure 1.48.

Why? The two triangles Δ ABK and ΔACT are equal due to the fact they each have an equal side (AC and AB are both the radii of the circle) and two equal angles. Also, $\widehat{AKB}$ equals $\widehat{ACT}$ since they both result from perpendicular lines BK and CT, respectively, and they equal 90°. Also, $\widehat{BAK}$ equals $\widehat{CAT}$ as they are both equal to θ. As a result, we can conclude the length of AT is equal to the length of AK.

Please note the angle θ we have chosen for both sine and cosine is between 0 and $\frac{\pi}{2}$ (90°). Also, be courteous how we solve the problem if the angle is in the third quadrant (for example sin (240 °)).

$cos(\theta) = length\ AK$

$cos(\pi - \theta) = -length\ of\ AT$

$$cos(\pi - \theta) = -cos(\theta)$$

Therefore, we can claim

$cos(120^o) = cos(180^o - 60^o) = -cos(60^o) = -\frac{1}{2}$

Examples:

Find the value of the sine for the following angles based on the common angles.

72. $\cos(150^o)$

Solution:

$$\cos(150^o) = \cos(180^o - 150^o) = -\cos(30^o) = -\frac{\sqrt{3}}{2}$$

---

73. $\cos(300^o)$

Solution:

Side note: Adding or subtracting 360° or 2π to the sine or cosine of an angle would not change the result as we get to the same point (period of both sine and cosine of an angle is 360° or 2π).

$$\cos(300^o) = \cos(300^o - 360^o) = \cos(-60^o) = \cos(60^o) = \frac{1}{2}$$

---

74. $\cos(330^o)$

Solution:

$$\cos(330^o) = \cos(330^o - 360^o) = \cos(-30^o) = \cos(30^o) = \frac{\sqrt{3}}{2}$$

---

75. $\cos(240^o)$

Solution:

$$\cos(240^o) = \cos(240^o - 360^o) = \cos(-120^o) = \cos(120^o)$$

$$= -\cos(180^o - 120^o) = -\cos(60^o) = -\frac{1}{2}$$

---

76. $\cos(225^o)$

Solution:

$$\cos(225^o) = \cos(225^o - 360^o) = \cos(-135^o) = \cos(135^o)$$

$$= -\cos(180^o - 135^o) = -\cos(45^o) = -\frac{\sqrt{2}}{2}$$

---

77. $\cos(\frac{4\pi}{3})$

Solution:

$$\cos\left(\frac{4\pi}{3}\right) = \cos\left(\frac{4\pi}{3} - 2\pi\right) = \cos\left(\frac{4\pi - 6\pi}{3}\right) = \cos\left(-\frac{2\pi}{3}\right) = \cos\left(\frac{2\pi}{3}\right)$$

$$= -\cos\left(\pi - \frac{2\pi}{3}\right) = -\cos\left(\frac{\pi}{3}\right) = -\frac{1}{2}$$

---

78. $\cos(-\frac{3\pi}{4})$

Solution:

$$\cos\left(\frac{-3\pi}{4}\right) = \cos\left(\frac{3\pi}{4}\right) = -\cos\left(\pi - \left(\frac{3\pi}{4}\right)\right) = -\cos\left(\frac{4\pi - 3\pi}{4}\right)$$

$$= -\cos\left(\frac{\pi}{4}\right) = -\frac{\sqrt{2}}{2}$$

---

2. Right Triangles

2.1 Height of a Building

I have a ruler that can measure up to 20 centimeters. I wanted to calculate the height of a nearby building. I counted the number of street lights (5) from where I am standing to the structure and knew the distance between the street light on my street (25 meters). I was then able to determine the distance from my position to the building, which was 125 meters. How could I measure is the height of the building?

Here is what I did. I lay down on the ground (I didn't have to!). Then I closed one eye and tried to keep my 20 cm (.2 meter) ruler perpendicular to the ground. I attempted to look in a way to construct a line, which passed my eye, the top of the ruler and the highest point of the building. Then I measured the distance from my eye to the bottom ruler using the same ruler (.25 meter). As a result, here is what I got pictorially.

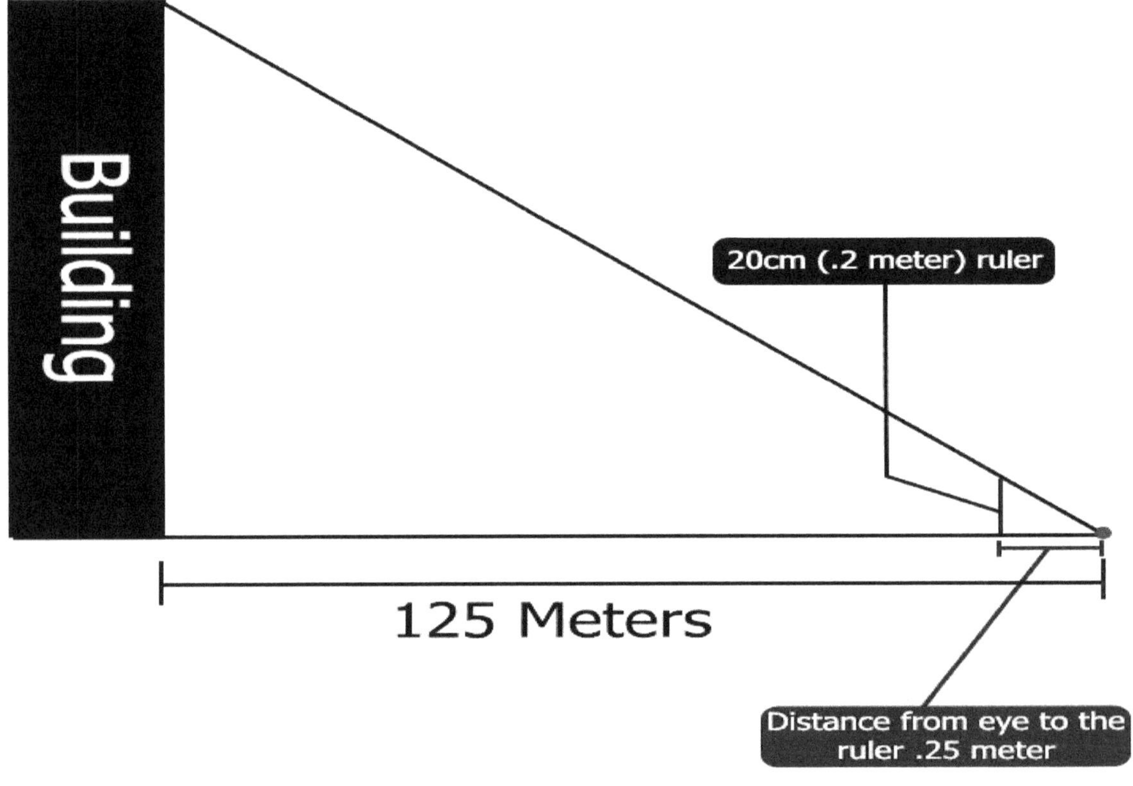

Figure 2.1

Can you see two triangles that we have constructed? You can spot a large triangle (ΔADE) and one smaller one (ΔABC) contained in the large one. Over 2,500 years ago, Thales noticed the triangles, and he understood they are similar (not necessarily equal but similar).

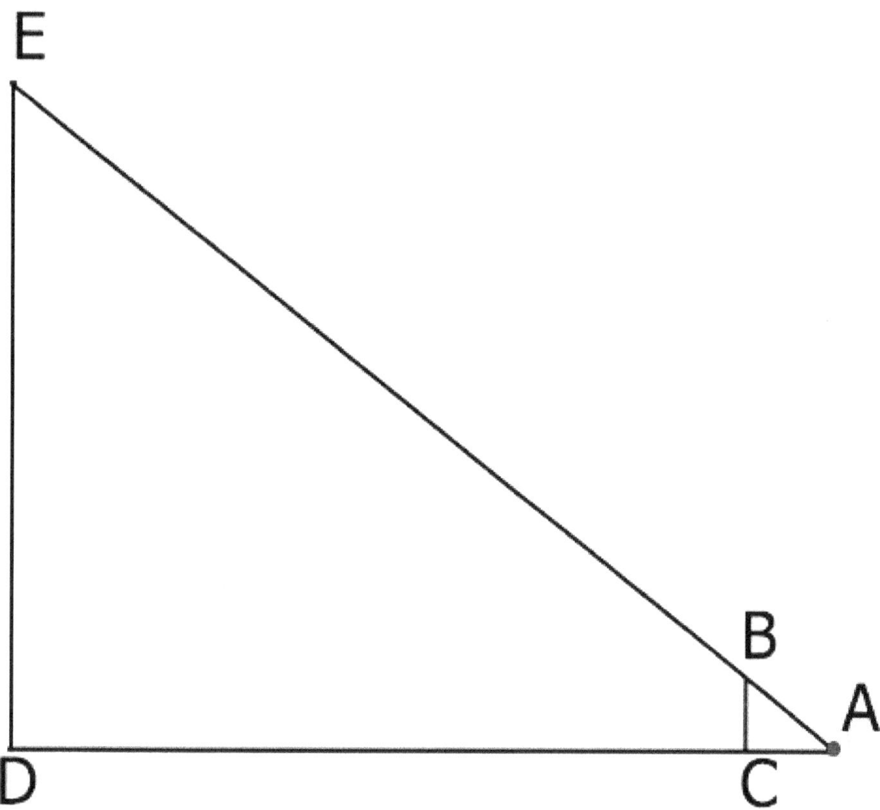

Figure 2.2.

The reason the two triangles are similar in our case is that they have three equal angles that make them similar. If you want to detect the angles, we have two normal lines, which would produce two 90^oangles ($\angle BCA = \angle EDA$) and a common angle ($\angle BAC = \angle EAD$). That is, two angles of each triangle are equal, and the last angle in each triangle must be equal as well due to the fact the total angles of a triangle is equal to 180°.

Note these triangles preserve the proportion. Therefore, the ratio of $\overline{\mathrm{AC}}$ (length of my eye to the ruler) to $\overline{\mathrm{AD}}$ (the distance of my eye to the building) is equal to $\overline{\mathrm{BC}}$ (the length of the ruler) to $\overline{\mathrm{BD}}$ (the height of the building). Therefore, we have the following:

What happens if you multiply both sides of an equation by zero?

You are right; we get zero on both sides. Please check if the following is correct:

$$2 \neq 9$$

I assume you replied yes (please:D). Now, if I multiply both sides by zero, the following would happen:

$$2.(0) = 9.(0) => \ 0 = 0$$

As you have noticed, I have multiplied both sides by zero, and I got an equality out of something that is not equal. The point that I want to make is if I multiply both sides by zero, I cannot make any conclusion as to whether the previous steps were correct or not. The aforementioned point may not be as obvious in the presented example, but you may encounter more significant problems while solving various math or real life problems.

$$\frac{\overline{\mathrm{AC}}}{\overline{\mathrm{AD}}} = \frac{\overline{\mathrm{BC}}}{\overline{\mathrm{ED}}(height\ of\ the\ building} => \frac{.25}{125} = \frac{.2}{\overline{\mathrm{ED}}}$$

We can now solve for $\overline{\mathrm{ED}}$ by multiplying both sides of the equality by $\overline{\mathrm{ED}}$ (keeping in mind $\overline{\mathrm{ED}}\ is\ not\ zero$) and dividing both sides by $\frac{.25}{150}$. As a result, we get the following equality:

$$\frac{.25}{125} = \frac{.2}{\overline{\mathrm{ED}}} => \overline{\mathrm{ED}}\ \frac{.25}{125} = \frac{.2}{\overline{\mathrm{ED}}}\ \overline{\mathrm{ED}} => \overline{\mathrm{ED}}\ \frac{.25}{125} = .2$$

$$\frac{\overline{ED}\ \frac{.25}{150}}{\frac{.25}{125}} = \frac{.2}{\frac{.25}{125}} => \overline{ED} = \frac{.2}{\frac{.25}{125}} => \overline{ED} = .2\ \frac{125}{.25} = 100$$

The height of the building is approximately 100 meters.

Examples:

Now let's find the length of the missing sides given the fact that triangles are similar.

1. Assume the two triangles have right angles. Investigate if they are similar and then find the value for "?".

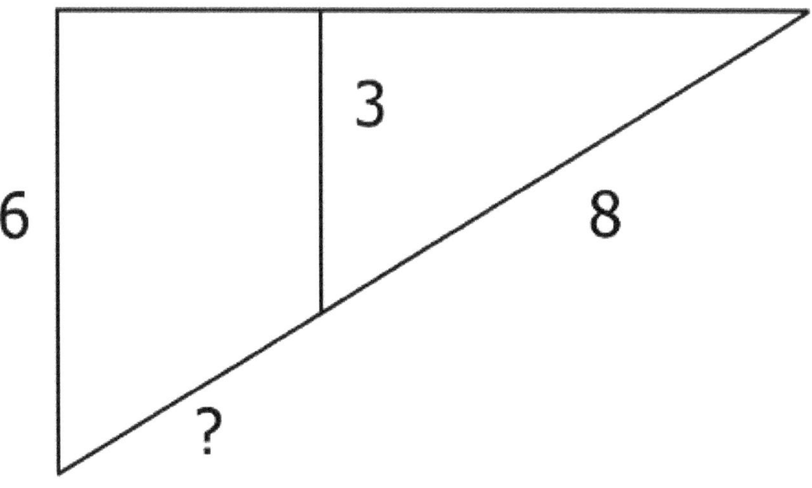

Figure 2.3.

To calculate "?" first, we need to make sure the two triangles are similar. They both have a right angle, and they share the same pick angles (the angle on the right on the graph). Since the total angles of a triangle are equal to 180 degrees, the last angles must also be equal.

The two triangles are similar; therefore, their ratio must be the same. That is, the base of the smaller triangle to the base of the larger triangle should have the same proportion as the hypotenuse of the smaller triangle to the larger triangle.

$\frac{3}{6} = \frac{8}{?+8} => 3 \times (?+8) = 6 \times 8$

$=> 3? + 24 = 48$

$=> 3? = 48 - 24 = 24 => ? = \frac{24}{3} = 8$

---

2. Given that the two triangles are similar, find "?". Thanks:D

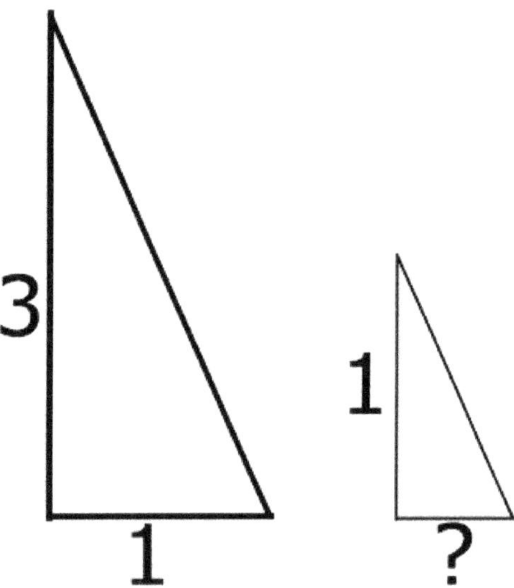

Figure 2.4.

All we have to do is to use the ratio to find "?".

$\frac{3}{1} = \frac{1}{?} => 1 \times 1 = 3 \times ?$

$=> 1 = 3? => ? = \frac{1}{3}$

3. Given that the two bases of the given triangles are parallel, investigate if the given triangles are similar. Then find "?".

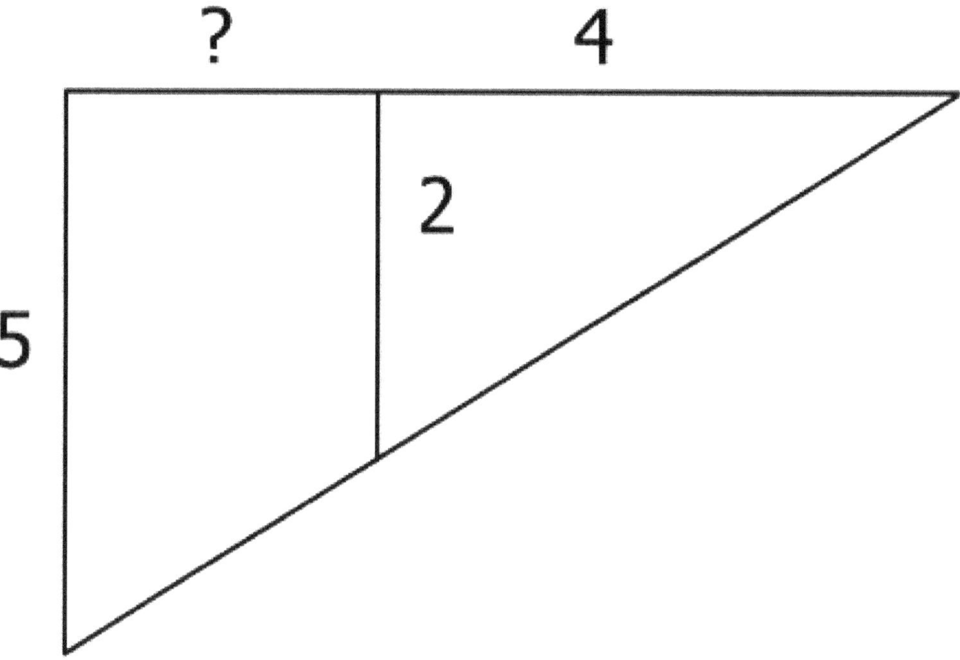

Figure 2.5.

Given that the two bases on the left are parallel, and the top line is passing through both, then the top line is a transversal line. This transversal causes the two obtained angles to have equal values.

Moreover, the common angle between the two triangles on the right side is another identical angle. The final angle in each triangle must be equivalent as the total angles of a triangle are equal. It is always sufficient to prove two angles are equal in two triangles if we are proving two triangles are similar. Now that we have established that the two triangles are similar, we can find "?" using the ratios.

$$\frac{2}{5} = \frac{4}{?+4} \Rightarrow 4 \times 5 = 2 \times (?+4)$$

$$\Rightarrow 20 = 2?+8 \Rightarrow 20 - 8 = 2?$$

$$\Rightarrow 12 = 2? \Rightarrow ? = \frac{12}{2} = 6$$

2.2 Properties of Equal Angles in Right Triangle

It should make sense that we can extend the sides of an angle as much as one prefers without changing the magnitude of the angle. That is, in the following examples, the magnitudes of the two angles are equal.

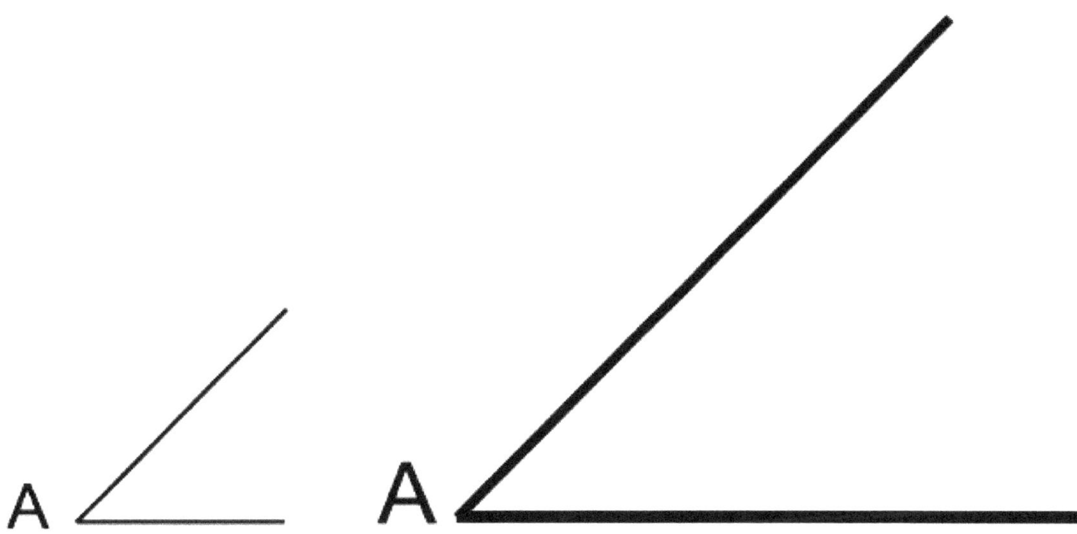

Figure 2.6.

Let's take this one step further. Try to create a right-angled triangle by connecting the two sides without changing the original angle. By doing so, you may obtain the following triangles.

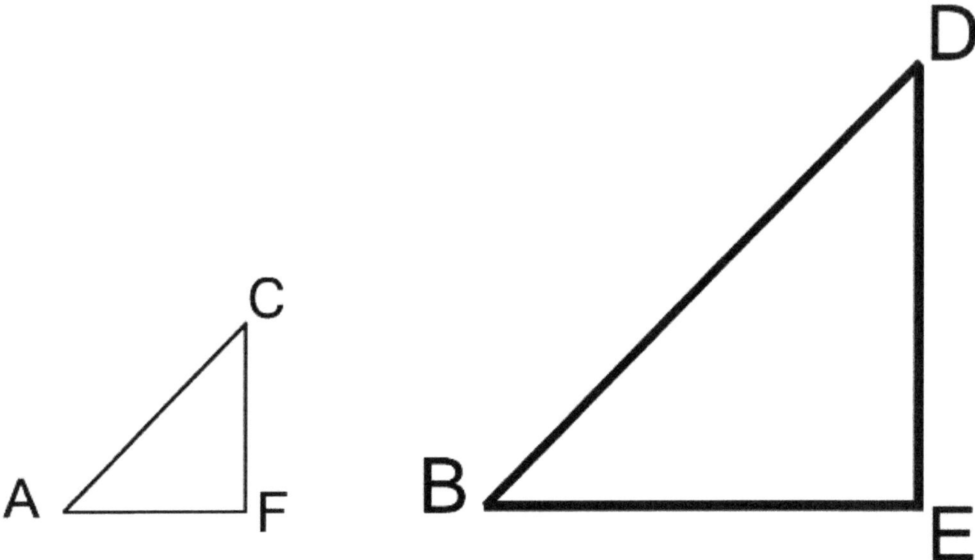

Figure 2.7.

Given ∠ A (angle A) = ∠ B and ∠ F = ∠ E = 90° we can conclude the following:

∠ A +∠ F + ∠ C =180° since Δ AFC (AFC is a triangle)

∠ B +∠ E + ∠ D =180° since Δ AFC (AFC is a triangle)

=> $\angle A + \angle F + \angle C = \angle B + \angle E + \angle D$

We replace ∠ A with ∠ B and ∠ F with ∠ E as we have established ∠ A = ∠ B and ∠ F = ∠ E.

$$\angle B + \angle E + \angle C = \angle B + \angle E + \angle D$$

After simplifying, we get the following:

$$\angle C = \angle D$$

One thing worth noting is if two right-angled triangles are equal (of course other than the 90° angle), they are similar (not necessarily equal).

We can relocate the two triangles and place them in a configuration analogous to the preceding one.

Notice sides FC and ED are parallel ($FC||ED$). According to the intercept theorem (Thales' theorem), the following ratios are obtained:

$$(\mathrm{E}-2.2.1)\frac{AF}{AE}=\frac{FC}{ED}=\frac{AC}{AD}$$

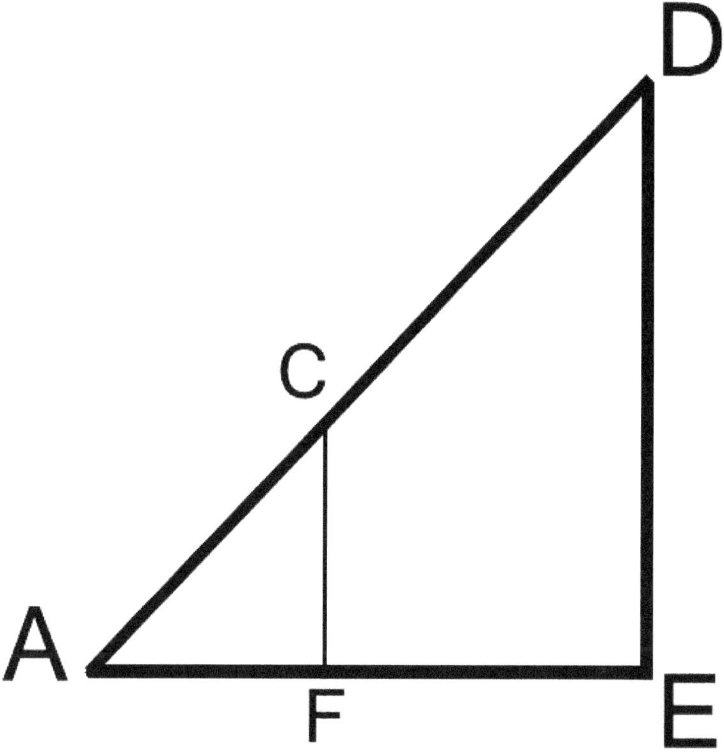

Figure 2.8.

As you can see, these ratios stay equal. Please note the angle ∠ A's role in the equalities. Now let's take one step further. Do you agree with the following?

$$\frac{FC}{AC}=\frac{ED}{AD}$$

From (E-2.2.1) ratio, it is easy to understand why. Here is one way to show why. Please check my work :D

We have already established the following:

$$\frac{FC}{ED} = \frac{AC}{AD}$$

Now multiply both sides by ED to get the following:

$$ED\frac{FC}{ED} = \frac{AC}{AD}ED => FC = \frac{AC}{AD}ED$$

Now divide both sides by AC to obtain the following:

$$\frac{FC}{AC} = \frac{ED}{AD}$$

That is, the smaller and larger triangles both have equal ratios. The factor that connects the two is the identical angle. We can further notice the resulting proportions:

$$\frac{AE}{ED} = \frac{AF}{FC} \qquad \frac{AE}{AD} = \frac{AF}{AC} \qquad \frac{ED}{AE} = \frac{FC}{AF}$$

$$\frac{AD}{ED} = \frac{AC}{FC} \qquad \frac{AD}{AE} = \frac{AC}{AF} \qquad \frac{ED}{AD} = \frac{FC}{AC}$$

We can further name these ratios associated with angles. Here we present an arbitrary triangle, and, for simplicity, we call the sides o (for opposite) a (for adjacent) and h (for hypotenuse) of $\angle$ A.

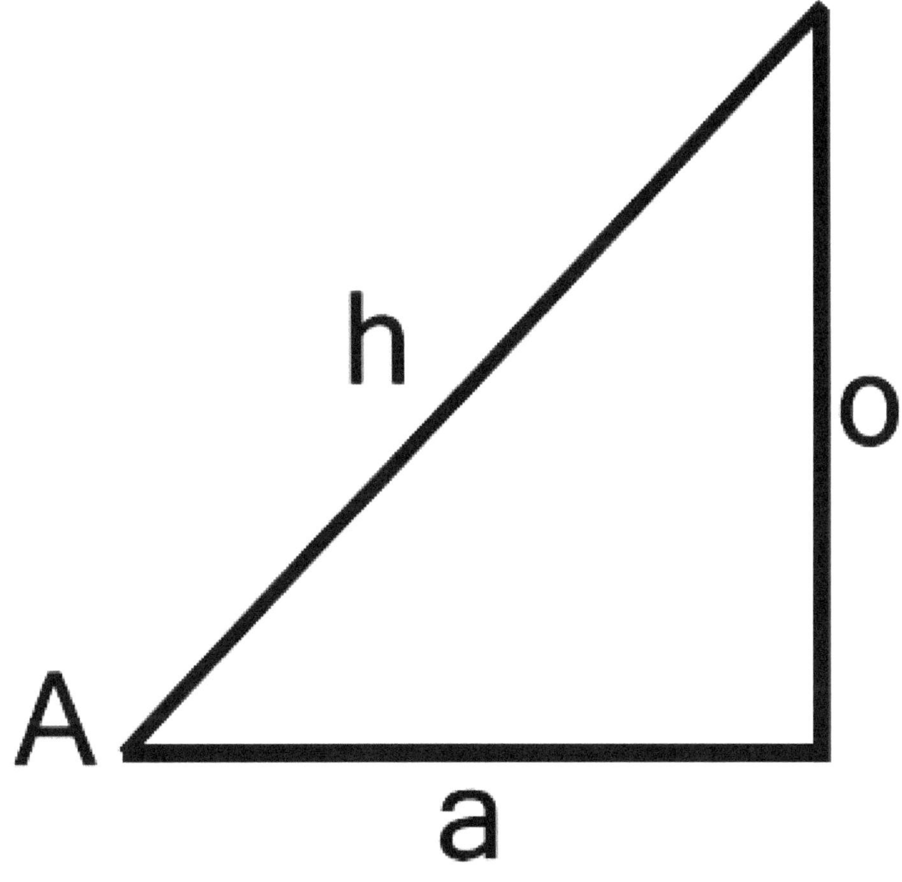

Figure 2.9.

$\frac{o}{a} = \tan(A)$ $\frac{h}{a} = \sec(A)$ $\frac{o}{h} = \sin(A)$

$\frac{h}{o} = \csc(A)$ $\frac{a}{o} = \cot(A)$ $\frac{a}{h} = \cos(A)$

Now let's establish Pythagorean relationships:

$$a^2 + o^2 = h^2$$

We can now divide both sides by h^2, and we obtain the following.

$$\frac{a^2}{h^2} + \frac{o^2}{h^2} = \frac{h^2}{h^2} => \frac{a^2}{h^2} + \frac{o^2}{h^2} = 1$$

Please note we named $\frac{a}{h}$ as $\cos(A)$ and $\frac{o}{h}$ as $\sin(A)$. Therefore, we can achieve the following:

$$\cos^2(A) + \sin^2(A) = 1$$

The derived equation is a notable result, and we will use it in many of the following problems.

2.3. Famous Right Triangles and Famous Ratios

Famous right triangles are triangles that possess one of the following two sets of angles. The set of 45°, 45° and 90° angles or the set of 30°, 60° and 90° angles. The aforementioned set of angles in a right triangle would lead to the ratios that we have shown in the famous angle section of the book in chapter one. Now you can see how we can explain the ratios.

45° -45° - 90° Triangle Trigonometric Ratios

Let's start with 45°, 45° and 90° triangles.

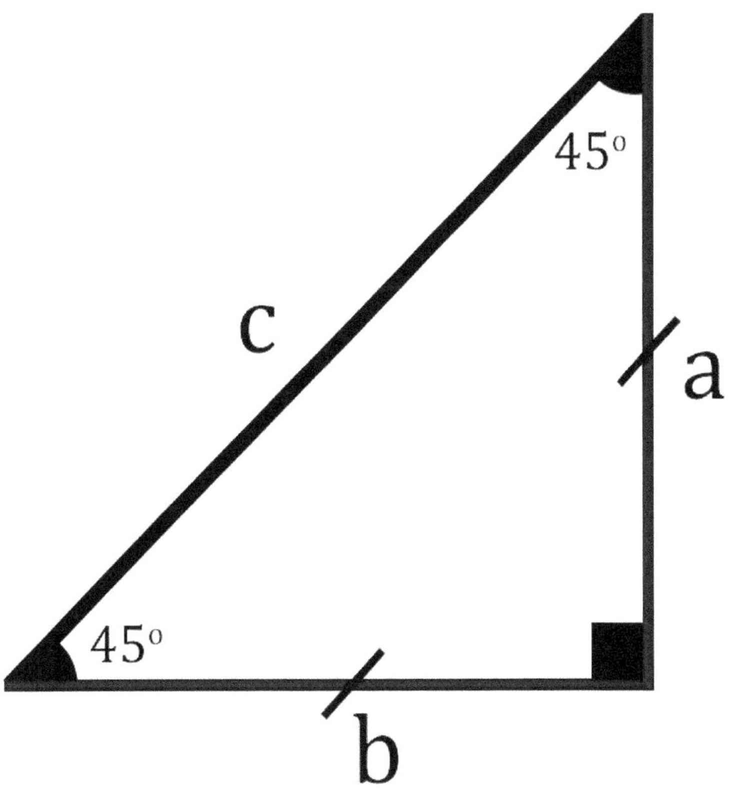

Figure 2.10.

Since we have two identical angles (45°), we can conclude that sides **a** and **b** are equal. Now let's apply this to the Pythagorean Theorem to define side **c** based on **a**.

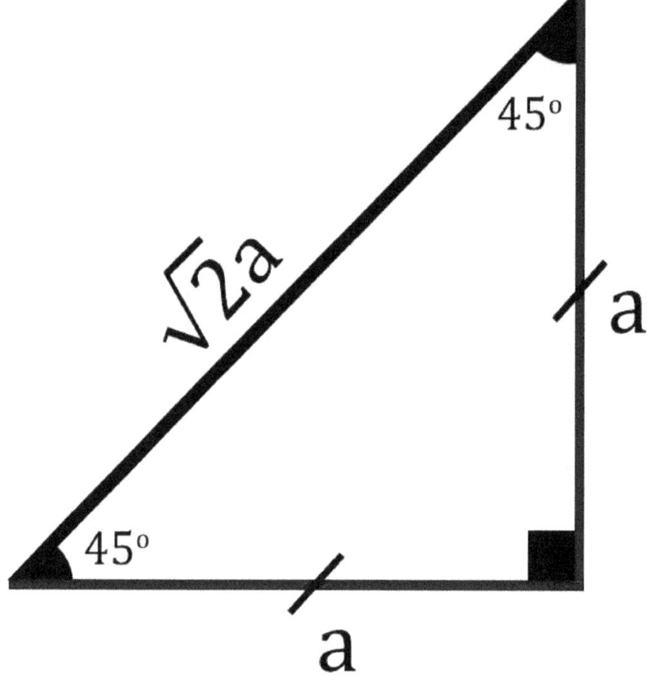

Figure 2.11.

$a^2 + b^2 = c^2$

Since $a = b \Rightarrow a^2 + a^2 = c^2$

$a^2 + a^2 = 2a^2 = c^2$

$\Rightarrow c = \sqrt{2a^2} = \sqrt{2}\sqrt{a^2} = \sqrt{2}a$

Based on our finding, let's rename the triangle sides as follows. Ok cool. Do you remember how we defined sin(x) and cos(x) for an angle? Here they are:

$$\sin(x) = \frac{Opposite}{Hypothenuse}$$

(Hypotenuse is the longest one :D)

$$\cos(x) = \frac{Adjacent}{Hypothenuse}$$

In our case, the angle x is 45°; therefore, we have the following:

$$\sin(45^o) = \frac{a}{\sqrt{2}a} = \frac{1}{\sqrt{2}}$$

To rationalize the denominator, we can multiply the numerator and denominator by $\sqrt{2}$, and we get the following value:

$$\sin(45^o) = \frac{1}{\sqrt{2}}.\frac{\sqrt{2}}{\sqrt{2}} = \frac{\sqrt{2}}{2}$$

$$\sin(45^o) = \frac{\sqrt{2}}{2}$$

Let's try it for **cosine**.

$$\cos(45^o) = \frac{a}{\sqrt{2}a} = \frac{1}{\sqrt{2}} = \frac{1}{\sqrt{2}}.\frac{\sqrt{2}}{\sqrt{2}} = \frac{\sqrt{2}}{2}$$

Subsequently, we calculated $\cos(45^o) = \frac{\sqrt{2}}{2}$

Is the result consistent with what we introduced earlier in the table? Long story short, we could determine the value without the knowledge of table.

The following results for other trigonometric ratios are easy to verify:

Tan(x):

$$\tan(45^o) = \frac{Opposite}{Adjecent} = \frac{a}{a} = 1$$

Or, equally well, you could say:

$$\tan(45^o) = \frac{\sin(45^o)}{\cos(45^o)} = \frac{\frac{\sqrt{2}}{2}}{\frac{\sqrt{2}}{2}} = 1$$

Cot(x):

$$\cot(45^o) = \frac{Adjecent}{Opposite} = \frac{a}{a} = 1$$

Equivalently, we could calculate cotangent as:

$$\cot(45^o) = \frac{\cos(45^o)}{\sin(45^o)} = \frac{\frac{\sqrt{2}}{2}}{\frac{\sqrt{2}}{2}} = 1$$

Sec(x):

$$\sec(45^o) = \frac{Hypothenuse}{Adjacent} = \frac{\sqrt{2}a}{a} = \sqrt{2}$$

Alternatively, we could say:

$$\sec(45^o) = \frac{1}{\cos(45^o)} = \frac{1}{\frac{1}{\sqrt{2}}} = \sqrt{2}$$

Csc(x):

$$\sec(45^o) = \frac{Hypothenuse}{opposite} = \frac{\sqrt{2}a}{a} = \sqrt{2}$$

We could also state:

$$\sec(45^o) = \frac{1}{\sin(45^o)} = \frac{1}{\frac{1}{\sqrt{2}}} = \sqrt{2}$$

As expected, all the results are consistent we what we have obtained from the table. We could also argue that, even if we had only one of the ratios for the triangles, we could verify the rest using the definitions of trigonometric ratio and the following identity:

$$sin^2(x) + cos^2(x) = 1$$

As an example, let's assume we obtain the value of the $\tan(45^o) = 1$

$\tan(45^o) = \frac{\sin(45^o)}{\cos(45^o)} = 1 =>$ Multiply both sides by $\cos(45^o)$ assuming $\cos(45^o)$ is not 0 (if it was then $\tan(45^o)$ would not have been 1:D).

$$\cos(45^o)\frac{\sin(45^o)}{\cos(45^o)} = 1\cos(45^o) => \sin(45^o) = \cos(45^o)$$

Now let's use $sin^2(x) + cos^2(x) = 1$ considering the fact $\sin(45^o) = \cos(45^o)$

$$sin^2(45^o) + cos^2(45^o) = sin^2(45^o) + sin^2(45^o) = 2sin^2(45^o) = 1$$

(Divide both sides by 2) => $sin^2(45^o) = \frac{1}{2}$

(Getting the square root of both sides as they are both positive)=> $sin(45^o) = \sqrt{\frac{1}{2}} = \frac{\sqrt{1}}{\sqrt{2}} = \frac{1}{\sqrt{2}}$

(If you like to rationalize the denominator, you can multiply the numerator and denominator by $\sqrt{2}$)

$$=> sin(45^o) = \frac{1}{\sqrt{2}}.\frac{\sqrt{2}}{\sqrt{2}} = \frac{\sqrt{2}}{2}$$

Since we have established $sin(45^o) = cos(45^o) => cos(45^o) = \frac{\sqrt{2}}{2}$

Accordingly, the following results are unavoidable:

Cot(x)

$$\cot(45^o) = \frac{\cos(45^o)}{\sin(45^o)} = \frac{\frac{\sqrt{2}}{2}}{\frac{\sqrt{2}}{2}} = 1$$

Otherwise, you could have argued:

$$\cot(45^o) = \frac{1}{\tan(45^o)} = \frac{1}{1} = 1$$

Sec(x)

$$\sec(45^o) = \frac{1}{\cos(45^o)} = \frac{1}{\frac{\sqrt{2}}{2}}$$

Multiply the denominator and numerator by $\sqrt{2}$ =>

$$\sec(45^o) = \frac{1}{\frac{\sqrt{2}}{2}} \cdot \frac{\sqrt{2}}{\sqrt{2}} = \frac{\frac{\sqrt{2}}{1}}{\frac{2}{2}} = \sqrt{2}$$

Csc(x)

$$\csc(45^o) = \frac{1}{\sin(45^o)} = \frac{1}{\frac{\sqrt{2}}{2}}$$

Multiply the denominator and numerator by $\sqrt{2}$ =>

$$\csc(45^o) = \frac{1}{\frac{\sqrt{2}}{2}} \cdot \frac{\sqrt{2}}{\sqrt{2}} = \frac{\frac{\sqrt{2}}{1}}{\frac{2}{2}} = \sqrt{2}$$

As expected! :D

Now we can investigate the 30°, 60°, and 90° triangle and find its ratios. Let's consider the following triangle to start:

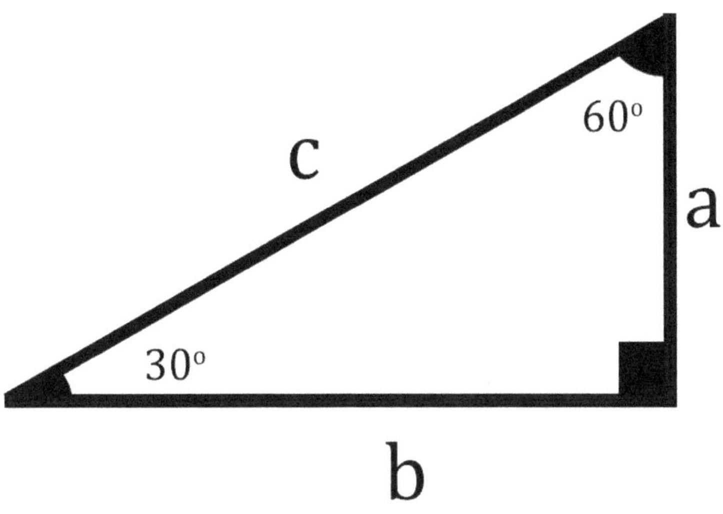

Figure 2.12.

There is more than one approach to proving this. See if you like the following path :D

First, let's draw a line "d" the same size as the side "a" on the side "c". The length of side "d" is equal to "a"; therefore, the angles must be equal. That is the angle made by side "d" and "a" is also 60°. The smaller angle resulted by the side "d" and "c" is 60° as well since the total angles in a triangle adds up to 180°. Therefore, we have an equilateral triangle.

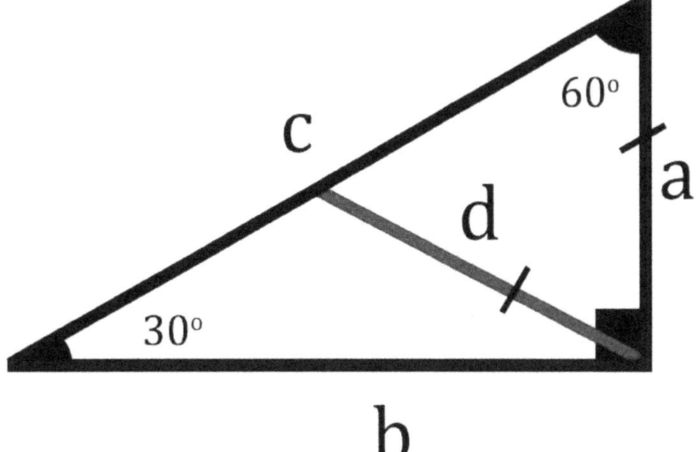

Figure 2.13.

On the other side, the angle resulted by the sides "d" and "b" must be 30° since the other part of the right angle is 60° (90°-60°=30°). Note the angle constructed by side "c" and side "b" is also 30°. Therefore, we have an isosceles triangle, and side "d" must be equal to side "c." Hence, we have the following equality between the length of the sides in the graph.

$$a = d = c_1 = c_2$$

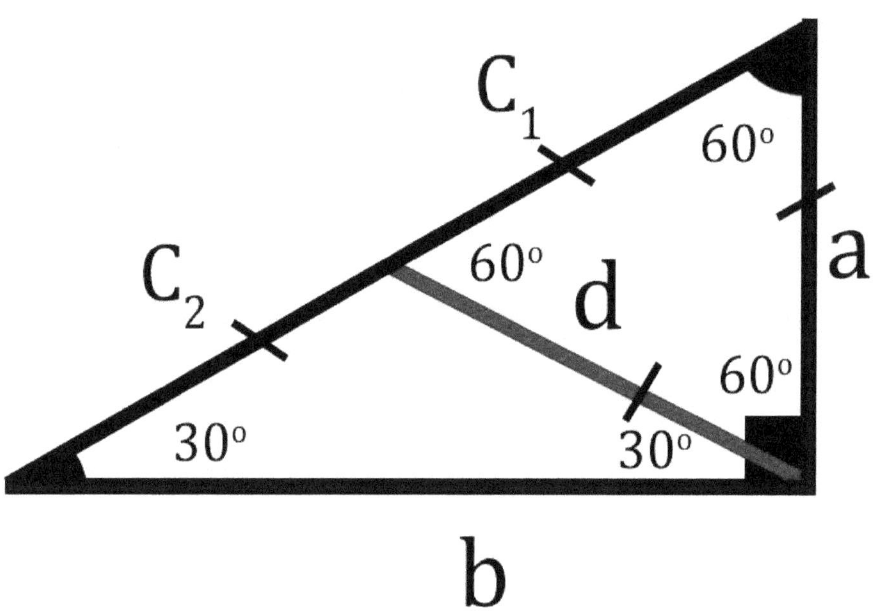

Figure 2.14.

More over the following are inavoidable.

$c_1 + c_2 = c => (c_1\ is\ equal\ to\ c_2)\ c_1 + c_1 = c$

$2c_1 = c => \ (c_1\ is\ equal\ to\ a) = 2a = c$

$b^2 + a^2 = c^2 = (2a)^2 = 4a^2 => b^2 + a^2 = 4a^2$

$=> (subtract\ a^2 from\ both\ sides)\ b^2 = 4a^2 - a^2 = 3a^2$

$$=> b = \sqrt{3a^2} = \sqrt{3}\sqrt{a^2} = \sqrt{3}a$$

Therefore, the graph can be represented as follows:

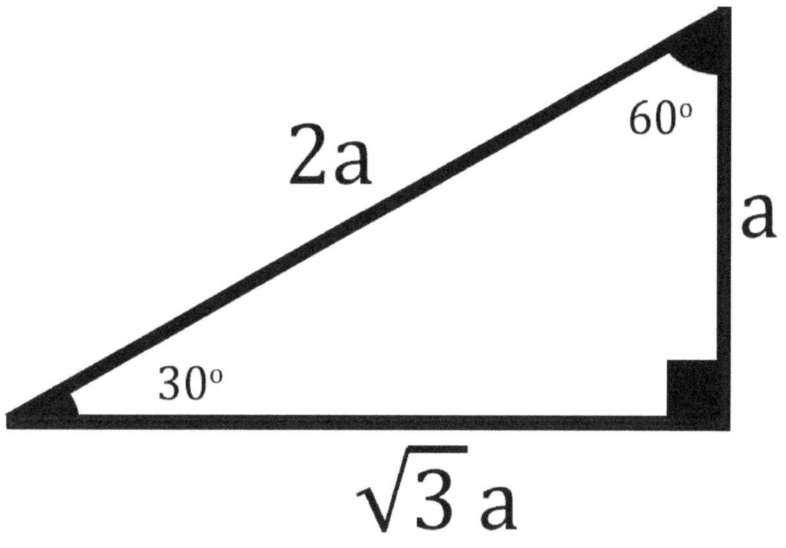

Figure 2.15.

Based on the last graph, let's see what the trigonometric ratio would be for a 30° angle.

Note for the 30° angle, we have $opposite = \mathrm{a}$, $hypotenuse = 2\mathrm{a}$, and $adjacent = \sqrt{3}\mathrm{a}$

$$\sin(x) = \frac{opposite}{hypotenuse} => \sin(30^o) = \frac{\mathrm{a}}{2\mathrm{a}} = \frac{1}{2}$$

$$\cos(x) = \frac{\text{adjacent}}{\text{hypotenuse}} => \cos(30^o) = \frac{\sqrt{3}\mathrm{a}}{2\mathrm{a}} = \frac{\sqrt{3}}{2}$$

$$\tan(x) = \frac{opposite}{\text{adjacent}} => \tan(30^o) = \frac{\mathrm{a}}{\sqrt{3}\mathrm{a}} = \frac{1}{\sqrt{3}} = \frac{1}{\sqrt{3}}\frac{\sqrt{3}}{\sqrt{3}} = \frac{\sqrt{3}}{3}$$

$$\cot(x) = \frac{1}{\tan(\mathrm{x})} => \cot(30^o) = \frac{1}{\frac{1}{\sqrt{3}}} = \sqrt{3}$$

$$\sec(x) = \frac{1}{\cos(x)} => \sec(30^o) = \frac{1}{\frac{\sqrt{3}}{2}} = \frac{2}{\sqrt{3}} = \frac{2}{\sqrt{3}}\frac{\sqrt{3}}{\sqrt{3}} = \frac{2\sqrt{3}}{3}$$

$$\csc(x) = \frac{1}{\sin(x)} => \csc(30^o) = \frac{1}{\frac{1}{2}} = 2$$

Now we use the same graph to investigate the 60° angle ratios.

Note for the 60° angle we have $opposite = \sqrt{3}\text{a}, hypotenuse = 2\text{a}$, and $adjacent = \text{a}$

$$\sin(x) = \frac{opposite}{hypotenuse} => \sin(60^o) = \frac{\sqrt{3}\text{a}}{2\text{a}} = \frac{\sqrt{3}}{2}$$

$$\cos(x) = \frac{\text{adjacent}}{\text{hypotenuse}} => \cos(60^o) = \frac{\text{a}}{2\text{a}} = \frac{1}{2}$$

$$\tan(x) = \frac{opposite}{\text{adjacent}} => \tan(60^o) = \frac{\sqrt{3}\text{a}}{\text{a}} = \sqrt{3}$$

$$\cot(x) = \frac{1}{\tan(x)} => \cot(60^o) = \frac{1}{\sqrt{3}} = \frac{1}{\sqrt{3}}\frac{\sqrt{3}}{\sqrt{3}} = \frac{\sqrt{3}}{3}$$

$$\sec(x) = \frac{1}{\cos(x)} => \sec(60^o) = \frac{1}{\frac{1}{2}} = 2$$

$$\csc(x) = \frac{1}{\sin(x)} => \csc(60^o) = \frac{1}{\frac{\sqrt{3}}{2}} = \frac{2}{\sqrt{3}} = \frac{2}{\sqrt{3}}\frac{\sqrt{3}}{\sqrt{3}} = \frac{2\sqrt{3}}{3}$$

Well, it seems we are consistent with the rest of the world as we know it :D

What we have discussed in this section for famous triangles rests on the angles; however, the special triangles could also be segregated based on their sides. Such division would lead to Pythagorean triples. The most famous ones are the 3-4-5 triples.

$\sqrt{3^2 + 4^2} = \sqrt{25} = 5$

There are many other ratios which can come from the multiples of the mentioned ratios. For example, 6-8-10. Let's check if I am right.

$\sqrt{6^2 + 8^2} = \sqrt{100} = 10$ (Of course it is not a proof!).

As we are just scratching the surface, I do not want to dive into the world of Pythagorean triples, but I just want to talk about an interesting formula by Euclid for generating Pythagorean triples. The formula states that the ratios of the sides must be in the form $m^2 - n^2$, $2mn$, and $m^2 + n^2$. Since we do not want to get any negative values for the length, we chose "m" to be an integer greater than integer "n." How can we check this? Well, using Pythagorean formula :D

$$a^2 + b^2 = c^2$$

Assume $a = m^2 - n^2, b = 2mn, c = m^2 + n^2$

$(m^2 - n^2)^2 + (2mn)^2 = m^4 - 2m^2n^2 + n^4 + 4m^2n^2 = m^4 + 2m^2n^2 + n^4$

$= (m^2 + n^2)^2 = c^2$

As we expected the value of c must be equal to $m^2 + n^2$

Just to get another base, you can choose m=3 and n=1 then the ratios must be:

$a = 3^2 - 1^2 = 8\,, b = 2(3)(1) = 6, c = 3^2 + 1^2 = 10 => 6 - 8 - 10$

Now it should make more sense why all the multiples of any Pythagorean triple would also be a Pythagorean triple. Let's say that if we multiply everything by integer km then we get the following:

Given $a^2 + b^2 = c^2$

Let's prove $(ka)^2 + (kb)^2 = (kc)^2$

$= k^2a^2 + k^2b^2 = k^2(a^2 + b^2) = k^2c^2 = (kc)^2$

$=> (ka)^2 + (kb)^2 = (kc)^2$

Now, you may ask, would all the Pythagorean triples be some multiples of 3-4-5 triples? The answer is not necessarily. Try m=3, and n=2, and we get the following triples:

$a = 3^2 - 2^2 = 5\,, b = 2(3)(2) = 12, c = 3^2 + 2^2 = 13 => 5, 12, 13$

I guess it is a good point to stop the discussion even though I would love to carry on.

3. Trigonometric Identities

3.1 Building from Scratch

Remember the unit circle? Radius 1? My assumption is your memory is similar to mine. In that case, you can see the circle as presented. Any point on the circle would have a sin(x) and a cos(x) value. Also, note that the radius of the circle is equal to one. Except for the points on the axes, you can construct a right-angled triangle using the point on the circle with the radius being the hypotenuse of the triangle.

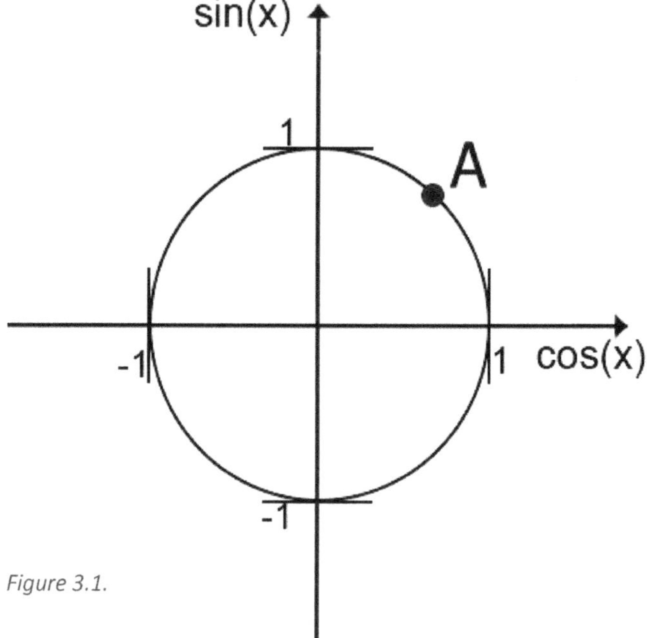

Figure 3.1.

As an example, take point A on the circle. Try to take the triangle out of the circle and see if you can find the Pythagorean relationship between its sides. By doing so, you would obtain the following shape. In the presented graph, it should be understandable why we can use the following formula:

$sin^2(x) + \cos^2(x) = 1^2$

Or simply:

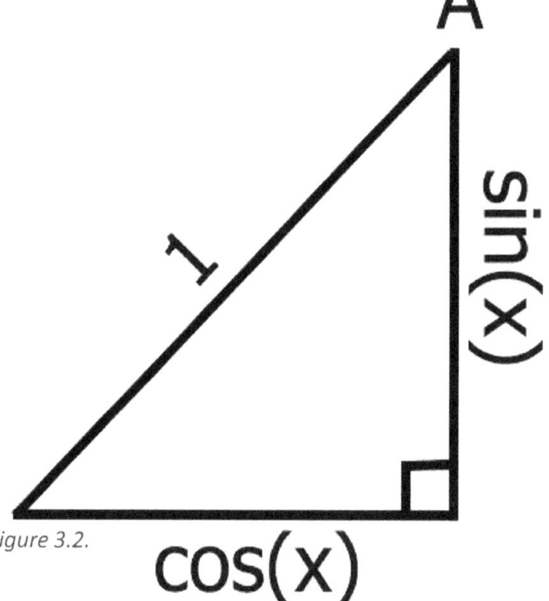

Figure 3.2.

$\boldsymbol{sin^2(x) + \cos^2(x) = 1}$ **(Identity 1)**

I would consider the aforementioned relationship as the building blocks of triangular identity. Feel free to go to 2.2. to revise another way to prove identity 1.

Since we have the identity, let's play a bit with it. Why don't we divide both sides of it by $sin^2(x)$ assuming $sin(x)$ is not equal to zero.

I believe it should make sense why we cannot divide anything by zero when solving an equation. No! Let's see why.

$\frac{12}{3} = 4$ We can check by saying $4 \times 3 = 12$

Let's try another one:

$\frac{32}{4} = 8$ Again, we can test by applying $8 \times 4 = 32$

How about the following?

$\frac{4}{0} =?$ Whatever you try it will not work. That is, nothing times zero would give you 4.

That is why we cannot divide by zero.

Here is what we get:

$$\frac{sin^2(x)+\cos^2(x)}{sin^2(x)} = \frac{1}{sin^2(x)} => \frac{sin^2(x)}{sin^2(x)} + \frac{\cos^2(x)}{sin^2(x)} = \frac{1}{sin^2(x)} = \csc^2(x)$$

which leads us to the next identity.

$\mathbf{1+\cot^2(x)=\csc^2(x)}$ **(Identity 2)**

Please recall the following relationships that we have defined previously:

$$\frac{\cos(x)}{\sin(x)} = \cot(x)$$

$$\frac{1}{\sin(x)} = \csc(x)$$

Now, why don't we divide both sides of identity 1 by $cos^2(x)$ assuming $cos(x)$ is not equal to zero. Here is what we get:

$$\frac{sin^2(x)+\cos^2(x)}{cos^2(x)} = \frac{1}{cos^2(x)} => \frac{sin^2(x)}{cos^2(x)} + \frac{\cos^2(x)}{cos^2(x)} = \frac{1}{cos^2(x)} = \sec^2(x)$$

Which can lead to the next identity:

$\mathbf{1+\tan^2(x)=\sec^2(x)}$ **(Identity 3)**

Please recall the following relationships that we have defined previously:

$$\frac{\sin(x)}{\cos(x)} = \tan(x)$$

$$\frac{1}{\cos(x)} = \sec(x)$$

$\boldsymbol{cos(-x) = \cos(x)}$ **(Identity 4)**

One way to investigate this identity is the following graph. Please note that, when we are talking about a positive angle, we refer to the counterclockwise rotation from the positive x-axis. Likewise, when we talk about negative angles, we are talking about the clockwise rotation starting from the positive x-axis. Note the length of AK is equal to cos(x). Please refer to the following figure to visualize identity 4:

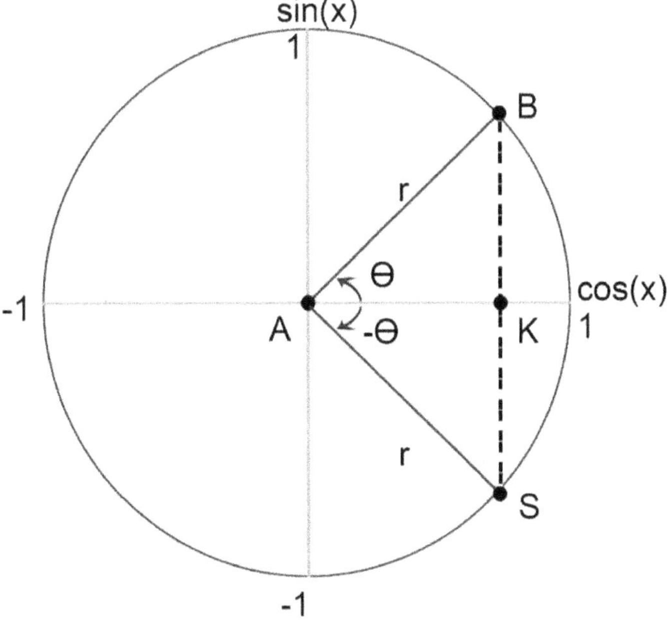

Figure 3.3.

$\boldsymbol{sin(-x) = -\sin(x)}$ **(Identity 5)**

Similar to the previous part, we can quickly illustrate Identity 5 using the following graph. It should be clear the length of AD is equal to sin(x), and the magnitude of AF is equal to sin(-x). Hence, we can observe AD and AF have the same length. The only difference is one stays on the positive side of the sin(x) axis while the other one lies on the negative part. Therefore, it can be concluded sin(-x)=-sin(x). Feel free to investigate the aforementioned equation using the following graph:

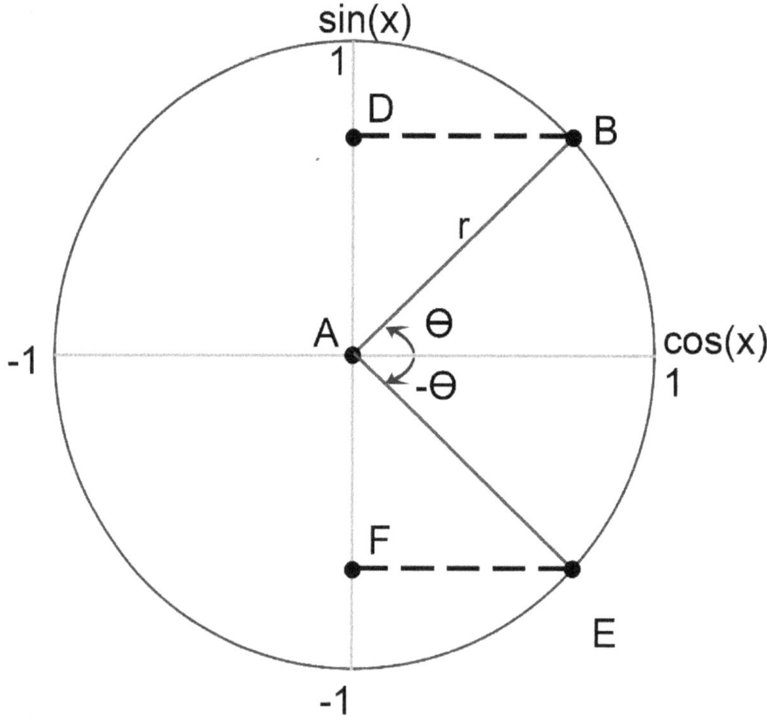

Figure 3.4.

$\boldsymbol{cos(x + 2\pi) = \cos(x)}$ **(Identity 6)**

6th identity is a direct consequence of the period of sin(x) and cos(x) functions. The period of both is 2π. That is, every 2π we get the same result for the sine and cosine function.

$sin(x + 2\pi) = \sin(x)$ **(Identity 7)**

7th identity is also a direct consequence of the period of sin(x) and cos(x) functions. The period of both is 2π. That is, every 2π we get the same result for the sine and cosine function. It is also valid to state the following identity for more general cases.

$cos(x + 2k\pi) = \cos(x)$ **(Identity 8)**

Where k is an integer (positive, negative, or zero). In mathematical terms, we can write $k \in \mathbb{Z}$.

8th identity is also a direct consequence of the period of sin(x) and cos(x) functions. The period of both is 2π. That is every 2π we get the same result for both the sine and cosine function.

$sin(x + 2k\pi) = \sin(x)$ **(Identity 9)**

Where k is an integer (positive, negative, or zero). In mathematical terms, we can write $k \in \mathbb{Z}$.

9th identity is also a direct consequence of the period of sin(x) and cos(x) functions. The period of both is 2π. That is, every 2π we get the same result for both the sine and cosine function.

3.2 Sin(x+y)

$$sin(x + y) = \sin(x)cos(y) + cos(x)sin(y) \qquad \textbf{(Identity 10)}$$

Before we start any explanation, please note we decided to show all the sides with lower case letters whereas we presented the angles and points with capital letters.

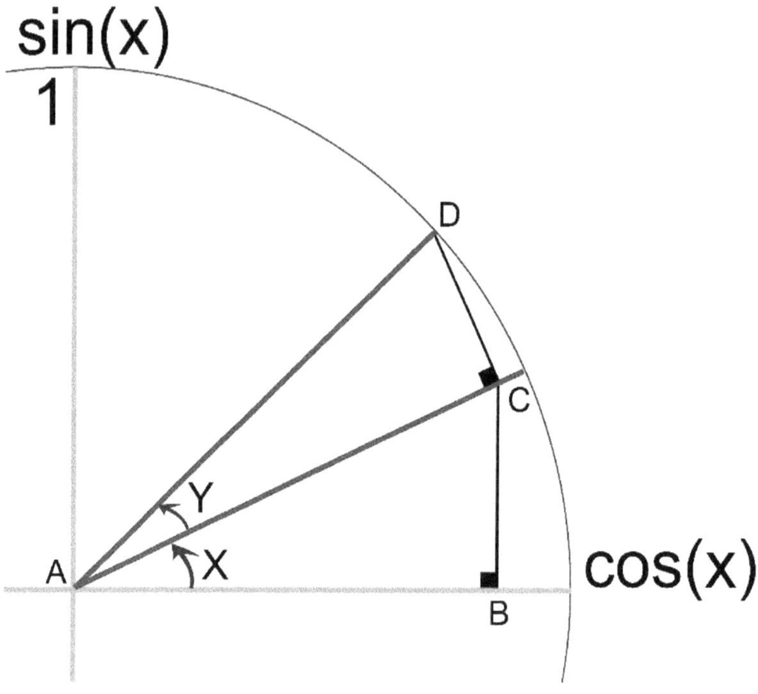

Figure 3.5.

Let's consider the following two triangles Δ ABC and Δ ACD. We want to investigate what sin(x+y) is.

To show the identity geometrically, let's draw an extra line as shown in the following graph and name the resulting angles and points. Note the line is perpendicular to cos (x) axis.

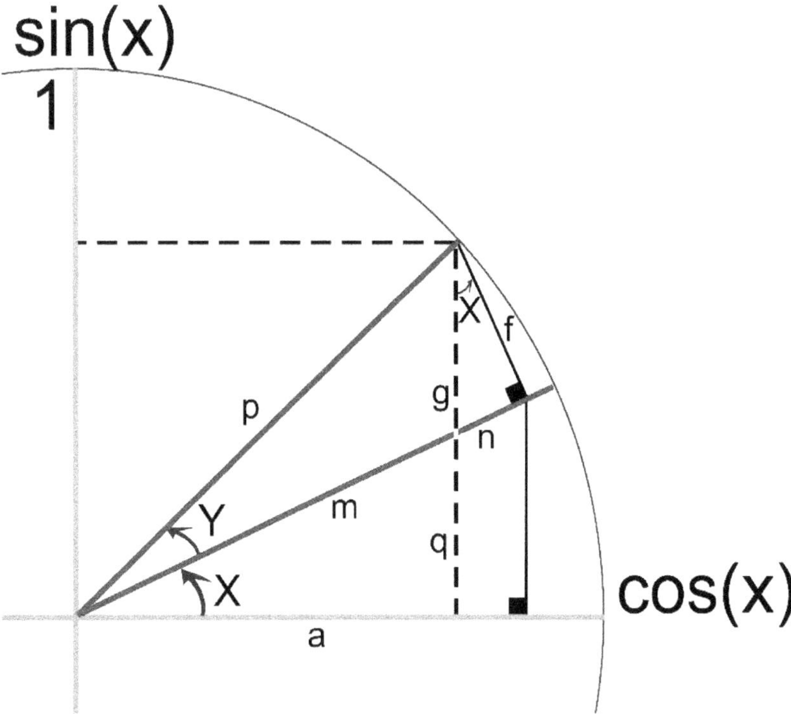

Figure 3.6.

Our goal is to find sin (x+y) in terms of angle ∠ X and ∠ Y. Geometrically, we can investigate the following:

$$\sin(\mathrm{x}+\mathrm{y})=\frac{g+q}{p}=\frac{g+q}{1}=g+q$$

Please check if the following ratios make sense:

$\sin(\mathrm{y})=\frac{f}{p}=\frac{f}{1}=>\sin(y)=f$ (i.10.1)

$\sin(\mathrm{x})=\frac{q}{m}$ (i.10.2)

$\sin(x)=\frac{n}{g}$ (i.10.3)

$n=g.sin(x)$ (i.10.4)

$$\cos(x) = \frac{a}{m} \quad \text{(i.10.5)}$$

$$\cos(x) = \frac{f}{g} \quad \text{(i.10.6)}$$

From (i.10.1) and (i.10.6)

$$\mathbf{g} = \frac{f}{\cos(x)} = \frac{\sin(y)}{\cos(x)} \quad \textbf{(i.10.7)}$$

$$\cos(y) = \frac{m+n}{p} = \frac{m+n}{1} = m+n \quad \text{(i.10.8)}$$

$$=> \cos(y) - n = m \quad \text{(i.10.9)}$$

From (i.10.2)

$$\mathrm{m.}\sin(\mathrm{x}) = q \quad \text{(i.10.10)}$$

From (i.10.4) and (i.10.7) (replace g by $\frac{\sin(y)}{\cos(x)}$)

$$n = \frac{\sin(y)}{\cos(x)}.\sin(x) = \frac{\sin(y).\sin(x)}{\cos(x)} \quad \text{(i.10.11)}$$

From (i.10.9) and (i.10.10) (replace g by $\cos(y) - n$)

$$(\cos(\mathrm{y}) - \mathrm{n}).\sin(\mathrm{x}) = q \quad \text{(i.10.12)}$$

$$(\cos(\mathrm{y}) - \frac{\sin(y).\sin(\mathrm{x})}{\cos(\mathrm{x})}).\sin(\mathrm{x}) = q \quad \text{(i.10.13)}$$

$$=> \mathbf{\cos(y)\sin(x)} - \frac{\mathbf{\sin(y).\sin^2(x)}}{\mathbf{\cos(x)}} = \boldsymbol{q} \quad \text{(i.10.14)}$$

We wanted to know what is q+p (in case you forgot why we are here :D)

From (i.10.7) and (i.10.14)

$$\mathrm{q} + \mathrm{g} = \cos(\mathrm{y})\sin(\mathrm{x}) - \frac{\sin(y).\sin^2(\mathrm{x})}{\cos(\mathrm{x})} + \frac{\sin(y)}{\cos(x)}$$

$= \cos(\mathrm{y})\sin(\mathrm{x}) + \frac{-\sin(y).\sin^2(\mathrm{x})+\sin(y)}{\cos(\mathrm{x})}$ (Let's factor out $\frac{\sin(y)}{\cos(\mathrm{x})}$ from the second term)

$$= \cos(y)\sin(x) + \frac{\sin(y)}{\cos(x)}(-\sin^2(x) + 1) \quad \text{(i.10.15)}$$

Using the first identity

$$cos^2(x) + \sin^2(x) = 1 \quad \text{(i.10.16)}$$

$$=> -\sin^2(x) + 1 = \cos^2(x) \quad \text{(i.10.17)}$$

As a result, (i.10.15) becomes:

$$\cos(y)\sin(x) + \frac{\sin(y)}{\cos(x)}\cos^2(x) \quad \text{(i.10.18)}$$

$$\cos(y).\sin(x) + \sin(y).\cos(x) \quad \text{(i.10.19)}$$

$$=> q + g = \cos(y).\sin(x) + \sin(y).\cos(x)$$

$$\sin(x + y) = \cos(y).\sin(x) + \sin(y).\cos(x)$$

Please note $\sin(x+y)$ is not equal to $\sin(x)+\sin(y)$. In fact, it is easy to show:

$$|\sin(x+y)| \leq |\sin(x)| + |\sin(y)|$$

Why?

$|\cos(x)| \leq 1$ multiply both sides by $|\sin(x)|$

$$\Rightarrow |\cos(x).\sin(x)| \leq |\sin(x)| \qquad (i.9.20)$$

Similarly, we have:

$|\cos(y)| \leq 1$ multiply both sides by $|\sin(y)|$

$$\Rightarrow |\cos(y).\sin(y)| \leq |\sin(y)| \qquad (i.9.21)$$

Now let's add left side of (i.9.20), the left side of (i.9.21) and the right side of (i.9.20) and the right side of (i.9.21) to obtain the following:

$$|\cos(x).\sin(x)| + |\cos(y).\sin(y)| \leq |\sin(x)| + |\sin(y)|$$

$$\Rightarrow |\cos(x).\sin(x) + \cos(y).\sin(y)| \leq |\cos(x).\sin(x)| + |\cos(y).\sin(y)| \leq |\sin(x)| + |\sin(y)|$$

Since $\sin(x+y) = \cos(y).\sin(x) + \sin(y).\cos(x)$

$$\Rightarrow |\sin(x+y)| \leq |\sin(x)| + |\sin(y)|$$

Done :D

3.3 Sin(y-x)

$\boldsymbol{sin(y-x) = \sin(y)cos(x) - cos(y)sin(x)}$ **(Identity 11)**

The proof is very similar to the previous part. Let's first understand the identity by visualizing it.

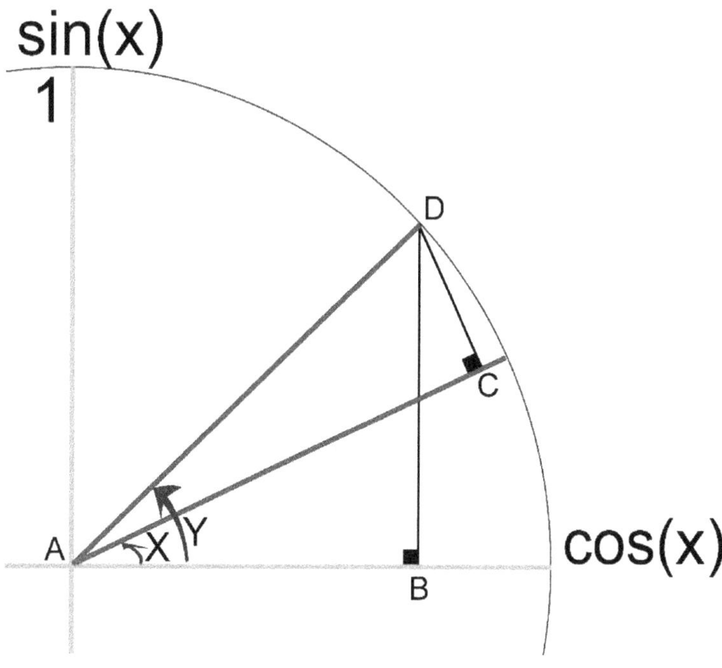

Figure 3.7.

As before, we save the lower case letters for sides and capital letters for angles and points. Let's consider the following two triangles, Δ ABD and Δ ACD. We want to investigate what sin(y-x) is.

Please try to find the reason why ∠x appears in two places.

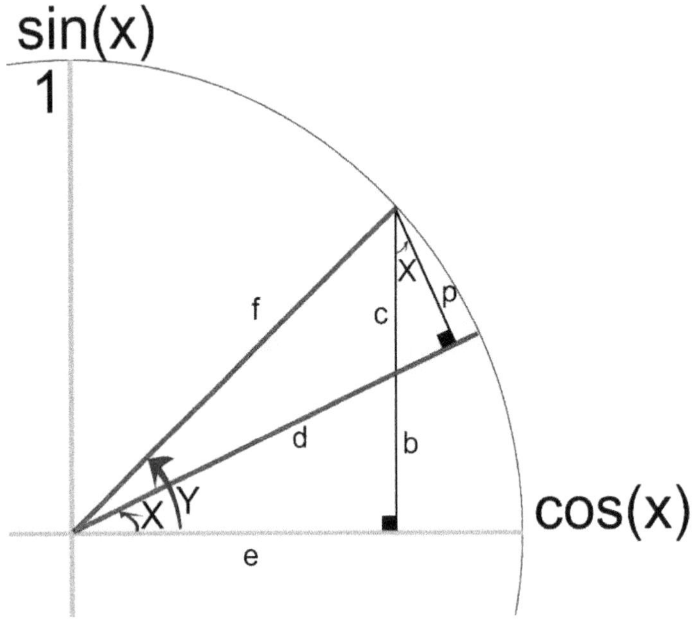

Figure 3.8.

Before anything, let's define our goal, which is the following:

$$\sin(y-x) = \frac{p}{f} = \frac{p}{1} \Rightarrow \sin(y-x) = p \qquad (i.11.1)$$

$$\sin(y) = \frac{b+c}{f} = \frac{b+c}{1} = b + c \qquad (i.11.2)$$

$$\cos(y) = \frac{e}{f} = e \qquad (i.11.3)$$

$$\sin(x) = \frac{b}{d} \Rightarrow d.\sin(x) = b \qquad (i.11.4)$$

$$\cos(x) = \frac{e}{d} \Rightarrow \frac{e}{\cos(x)} = d \qquad (i.11.5)$$

From (i.11.3) and (i.11.5) we can conclude the following:

$$d = \frac{\cos(y)}{\cos(x)} \qquad (i.11.6)$$

$\cos(x) = \frac{p}{c}$ (note the smaller triangle) (i.11.7)

=> $\boldsymbol{p = c.\cos(x)}$ **(i.11.8)**

From (i.11.4) and (i.11.6)

$$b = \frac{\cos(y)}{\cos(x)}\sin(x) \quad \text{(i.11.9)}$$

From (i.11.2)

$\sin(y) = b + c$ (i.11.10)

=> $\sin(y) - b = c$ (i.11.11)

From (i.11.9) and (i.11.11)

$$\sin(y) - \frac{\cos(y)}{\cos(x)}\sin(x) = \sin(y) - \frac{\cos(y).\sin(x)}{\cos(x)} =$$

$$= \frac{\sin(y)\cos(x) - \cos(y).\sin(x)}{\cos(x)} \quad \text{(i.11.12)}$$

From (i.11.11)

$$c = \frac{\sin(y)\cos(x) - \cos(y).\sin(x)}{\cos(x)} \quad \text{(i.11.13)}$$

From (i.11.8) and (i.11.13)

$$p = \left(\frac{\sin(y)\cos(x) - \cos(y).\sin(x)}{\cos(x)}\right)\cos(x)$$

$p = \sin(y)\cos(x) - \cos(y).\sin(x)$ (i.11.14)

=> $\sin(y - x) = \sin(y)\cos(x) - \cos(y).\sin(x)$

It is finished, and, yes, we could have written the following :D

$\sin(x - y) = \sin(x)\cos(y) - \cos(x).\sin(y)$

3.4 cos(x+y)

$cos(x + y) = \cos(x)cos(y) - sin(x)sin(y)$ **(Identity 12)**

As before, we save the lower case letters for sides and capital letters for angles and points. Let's consider the following two triangles, Δ ABC and Δ ACD. We want to investigate what cos(x+y) is.

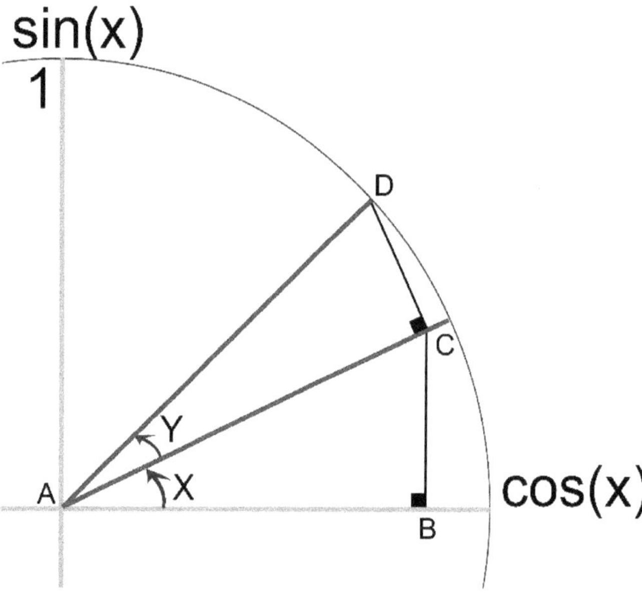

Figure 3.9.

Please try to find the reason why ∠x appears in two places.

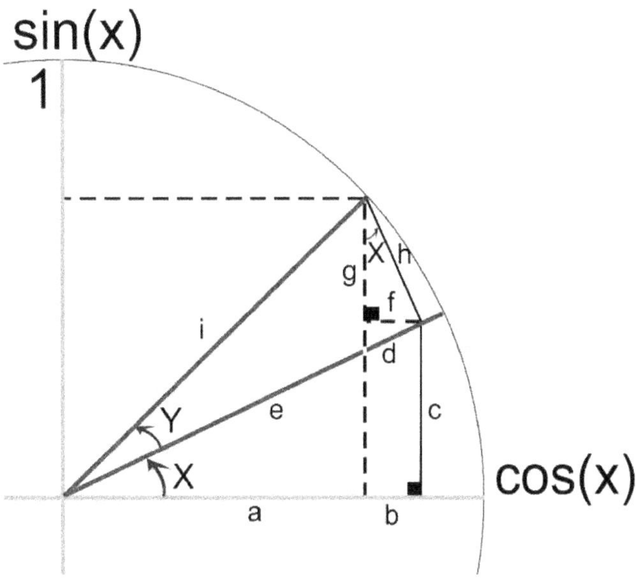

Figure 3.10.

Before anything, let's define our goal, which is the following:

$$\cos(\mathrm{x}+\mathrm{y}) = \frac{a}{i} = \frac{a}{1} => \cos(x+y) = a \qquad \text{(i.12.1)}$$

$$\cos(\mathrm{x}) = \frac{a+b}{e+d} \qquad \text{(i.12.2)}$$

$$\sin(x) = \frac{c}{e+d} \qquad \text{(i.12.3)}$$

$$sin(y) = \frac{h}{i} = \frac{h}{1} => \sin(y) = h \qquad \text{(i.12.4)}$$

$$\cos(y) = \frac{e+d}{i} = \frac{e+d}{1} => \cos(y) = e+d \qquad \text{(i.12.5)}$$

From (i.12.2) and (i.12.5), we can conclude the following:

$$\cos(\mathrm{x}) = \frac{\mathrm{a+b}}{\cos(y)} \qquad \text{(i.12.6)}$$

$$=> \cos(\mathrm{x}) . \cos(\mathrm{y}) = \mathrm{a} + \mathrm{b} \qquad \text{(i.12.7)}$$

$=> \cos(x).\cos(y) - b = a$ (i.12.8)

From (i.12.3) and (i.12.5), we can conclude the following:

$=> \sin(x) = \frac{c}{\cos(y)}$ (i.12.9)

From the smaller triangle, we can define the following:

$\sin(x) = \frac{f}{h} => h.\sin(x) = f$ (i.12.10)

From (i.12.4) and (i.12.10):

$\sin(y)\sin(x) = f$ (i.12.11)

$f = b$ (i.12.12)

From (i.12.8) and (i.12.12):

$=> \cos(x).\cos(y) - f = a$ (i.12.12)

$=> \cos(x).\cos(y) - \sin(y)\sin(x) = a$ (i.12.13)

$=> \cos(x + y) = \cos(x).\cos(y) - \sin(y)\sin(x)$

:D Yes, we are done.

3.5 cos(x-y)

$\boldsymbol{cos(x - y) = \cos(x)cos(y) + sin(x)sin(y)}$ **(Identity 13)**

As before, we save the lower case letters for sides and capital letters for angles and points. Let's consider the following two triangles, Δ ABD and Δ ACD. We want to investigate what is cos(x-y).

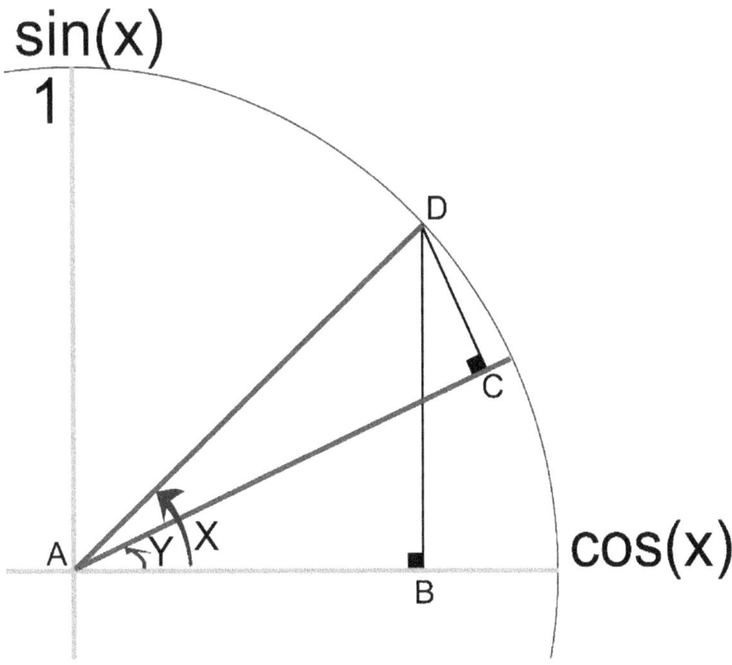

Figure 3.11.

Please try to find the reason why ∠y appears in two places.

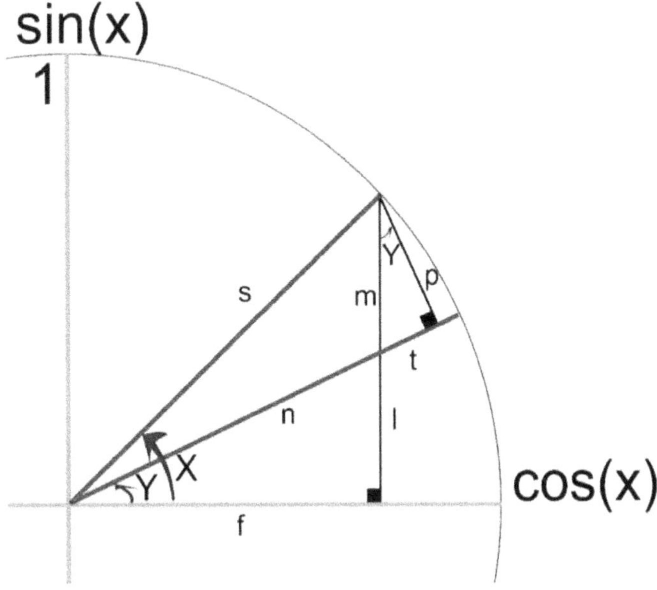

Figure 3.12.

Before anything, let's define our goal, which is the following:

$$\cos(x-y)=\frac{n+t}{s}=\frac{n+t}{1}$$

$=> \cos(x-y)=n+t$ (i.13.1)

$\sin(x)=\frac{l+m}{s}=\frac{l+m}{1}=l+m$ (i.13.2)

$=> m=\sin(x)-l$ (i.13.3)

$\cos(x)=\frac{f}{s}=\frac{f}{1}=f$ (i.13.4)

$sin(y)=\frac{l}{n}$ (i.13.5)

$sin(y)=\frac{t}{m}$ (from smaller triangle) (i.13.6)

$=> t=m.\sin(y)$ (i.13.7)

$\cos(y)=\frac{f}{n}$ (i.13.8)

From (i.13.3) and (i.13.7), we can conclude the following:

$t=(\sin(x)-l).\sin(y)$ (i.13.9)

From (i.13.5)

$l=n.\sin(y)$ (i.13.10)

From (i.13.4) and (i.13.8)

$n=\frac{f}{\cos(y)}=\frac{\cos(x)}{\cos(y)}$ (i.13.11)

From (i.13.10) and (i.13.11)

$l=\frac{\cos(x)}{\cos(y)}.\sin(y)$ (i.13.12)

From (i.13.9) and (i.13.12)

$$t = (\sin(x) - \frac{\cos(x)}{\cos(y)} . \sin(y)) . \sin(y) \qquad \text{(i.13.13)}$$

From (i.13.11) and (i.13.13)

$$n + t = \frac{\cos(x)}{\cos(y)} + (\sin(x) - \frac{\cos(x)}{\cos(y)} . \sin(y)) . \sin(y)$$

$$= \frac{\cos(x)}{\cos(y)} + \sin(x) . \sin(y) - \frac{\cos(x) . \sin^2(y)}{\cos(y)}$$

$$= \sin(x) . \sin(y) + \frac{\cos(x)}{\cos(y)} - \frac{\cos(x) . \sin^2(y)}{\cos(y)}$$

$$= \sin(x) . \sin(y) + \frac{\cos(x) - \cos(x) . \sin^2(y)}{\cos(y)}$$

Factor out cos(x) from the numerator of the second term

$$= \sin(x) . \sin(y) + \frac{\cos(x)(1 - \sin^2(y))}{\cos(y)} \qquad \text{(i.13.14)}$$

From $\sin^2(y) + \cos^2(y) = 1$

$$=> \cos^2(y) = 1 - \sin^2(y) \qquad \text{(i.13.15)}$$

From (i.13.14) and (i.13.15)

$$n + t = \sin(x) . \sin(y) + \frac{\cos(x) \cos^2(y)}{\cos(y)}$$

Cancel cos(y) in the second term.

$$= \sin(x) \sin(y) + \cos(x) \cos(y) \qquad \text{(i.13.16)}$$

$$\cos(x - y) = \sin(x) \sin(y) + \cos(x) \cos(y)$$

We are done :D and, yes, this is also equal to the following:

$$\cos(x - y) = \cos(\mathrm{x}) \, cos(y) + \sin(\mathrm{x}) \sin(y)$$

$$\cos(y - x) = \cos(\mathrm{x}) \, cos(y) + \sin(\mathrm{x}) \sin(y)$$

$$\cos(y - x) = \sin(\mathrm{x}) \sin(y) + \cos(\mathrm{x}) \, cos(y)$$

How to check quickly? Refer to identity 4 where we established

$\cos(x) = \cos(-x)$

Therefore, the following should also make sense:

$\cos(x - y) = \cos(-(x - y)) = \cos(-x + y) = \cos(y - x)$

3.6 Other identities

$\mathbf{tan}(\boldsymbol{x}+\boldsymbol{y}) = \frac{\mathbf{tan(x)+tan(y)}}{\mathbf{1-tan}(\boldsymbol{x})\mathbf{.tan}(\boldsymbol{y})}$ **(Identity 14)**

Proof:

By definition, we have the following:

$$\tan(x+y) = \frac{\sin(x+y)}{\cos(x+y)}$$

From previous identities, we have:

$$\sin(x+y) = \sin(x)\cos(y) + \sin(y)\cos(x)$$

$$\cos(x+y) = \cos(x)\cos(y) - \sin(x)\sin(y)$$

Therefore, the following can be achieved:

$$\tan(x+y) = \frac{\sin(x)\cos(y) + \sin(y)\cos(x)}{\cos(x)\cos(y) - \sin(x)\sin(y)}$$

Divide numerator and denominator by cos(x).cos(y)

$$\tan(x+y) = \frac{\frac{\sin(x)\cos(y)}{\cos(x)\cos(y)} + \frac{\sin(y)\cos(x)}{\cos(x)\cos(y)}}{\frac{\cos(x)\cos(y)}{\cos(x)\cos(y)} - \frac{\sin(x)\sin(y)}{\cos(x)\cos(y)}}$$

$$= \frac{\frac{\sin(x)}{\cos(x)} + \frac{\sin(y)}{\cos(y)}}{1 - \frac{\sin(x)\sin(y)}{\cos(x)\cos(y)}} = \frac{\tan(x) + \tan(\mathrm{y})}{1 - \tan(\mathrm{x})\tan(\mathrm{y})}$$

$$\mathbf{tan}(x-y)=\frac{\mathbf{tan(x)-tan(y)}}{\mathbf{1+tan}(x)\mathbf{.tan}(y)}$$ **(Identity 15)**

Proof:

By definition, we have the following:

$$\tan(x-y)=\frac{\sin(x-y)}{\cos(x-y)}$$

From previous identities, we have:

$$\sin(x-y)=\sin(x)\cos(y)-\sin(y)\cos(x)$$

$$\cos(x-y)=\cos(x)\cos(y)+\sin(x)\sin(y)$$

Therefore, the following is true:

$$\tan(x-y)=\frac{\sin(x)\cos(y)-\sin(y)\cos(x)}{\cos(x)\cos(y)+\sin(x)\sin(y)}$$

Divide numerator and denominator by cos(x)cos(y)

$$\tan(x-y)=\frac{\frac{\sin(x)\cos(y)}{\cos(x)\cos(y)}-\frac{\sin(y)\cos(x)}{\cos(x)\cos(y)}}{\frac{\cos(x)\cos(y)}{\cos(x)\cos(y)}+\frac{\sin(x)\sin(y)}{\cos(x)\cos(y)}}$$

$$=\frac{\frac{\sin(x)}{\cos(x)}-\frac{\sin(y)}{\cos(y)}}{1+\frac{\sin(x)\sin(y)}{\cos(x)\cos(y)}}=\frac{\tan(x)-\tan(y)}{1+\tan(x)\tan(y)}$$

$$\mathbf{cot}(\boldsymbol{x}+\boldsymbol{y}) = \frac{\mathbf{1-tan}(\boldsymbol{x}).\mathbf{tan}(\boldsymbol{y})}{\mathbf{tan(x)+tan(y)}}$$ **(Identity 16)**

Proof:

By definition, we have the following:

$$\cot(x+y) = \frac{1}{\tan(x+y)}$$

From previous identities, we have:

$$\tan(x+y) = \frac{\tan(x)+\tan(\mathrm{y})}{1-\tan(\mathrm{x})\tan(\mathrm{y})}$$

Just flip the result for tan(x-y) to get cot (x-y)

$$\cot(x+y) = \frac{1-\tan(x).\tan(y)}{\tan(\mathrm{x})+\tan(\mathrm{y})}$$

$$\mathbf{cot}(\boldsymbol{x}-\boldsymbol{y}) = \frac{\mathbf{1+tan}(\boldsymbol{x}).\mathbf{tan}(\boldsymbol{y})}{\mathbf{tan(x)-tan(y)}}$$ **(Identity 17)**

Proof:

By definition, we have the following:

$$\cot(x-y) = \frac{1}{\tan(x-y)}$$

From previous identities, we have:

$$\tan(x-y)=\frac{\tan(x)-\tan(y)}{1+\tan(x)\tan(y)}$$

Just flip the result for tan(x-y) to get cot (x-y)

$$\cot(x-y)=\frac{1+\tan(x).\tan(y)}{\tan(x)-\tan(y)}$$

$$\mathbf{\sin(2x)=2.\sin(x).\cos(x)}$$ **(Identity 18)**

Proof:

From

$$\sin(x+y)=\sin(x)\cos(y)+\sin(y)\cos(x)$$

We can achieve:

$$\sin(x+x)=\sin(x)\cos(x)+\sin(x)\cos(x)$$

$$=>\sin(2x)=2.\sin(x).\cos(x)$$

$$\mathbf{\cos(2x)=\cos^2(x)-\sin^2(x)}$$ **(Identity 19)**

From

$$\cos(x+y)=\cos(x)\cos(y)-\sin(x)\sin(y)$$

We can achieve:

$\cos(x + x) = \cos(x)\cos(x) - \sin(x)\sin(x)$

$=> \cos(2x) = \cos^2(x) - \sin^2(x)$

$\cos(2\mathrm{x}) = 2.\cos^2(x) - 1$ **(Identity 20)**

From

$\cos(x + y) = \cos(x)\cos(y) - \sin(x)\sin(y)$

We can achieve:

$\cos(x + x) = \cos(x)\cos(x) - \sin(x)\sin(x)$

$=> \cos(2x) = \cos^2(x) - \sin^2(x)$

Replace $-\sin^2(x)$ with $\cos^2(x) - 1$ (by rearranging the identity

$\sin^2(x) + \cos^2(x) = 1)$

$=> \cos(2x) = \cos^2(x) + \cos^2(x) - 1$

$=> \cos(2x) = 2.\cos^2(x) - 1$

$\cos(2\mathrm{x}) = 1 - 2.\sin^2(x)$ **(Identity 21)**

From

$\cos(x + y) = \cos(x)\cos(y) - \sin(x)\sin(y)$

We can achieve:

$\cos(x+x) = \cos(x)\cos(x) - \sin(x)\sin(x)$

$=> \cos(2x) = \cos^2(x) - \sin^2(x)$

Replace $\cos^2(x)$ with $1 - \sin^2(x)$ (rearranging the identity)

$(\sin^2(x) + \cos^2(x) = 1)$

$=> \cos(2x) = 1 - \sin^2(x) \; - \sin^2(x)$

$=> \cos(2x) = 1 - 2.\sin^2(x)$

$\mathbf{\cos^2(x) = \frac{1+\cos(2x)}{2}}$ **(Identity 22)**

From

$\cos(2\text{x}) = 2.\cos^2(x) - 1$

$=> \cos(2\text{x}) + 1 = 2.\cos^2(x)$

$=> \dfrac{\cos(2x)+1}{2} = \cos^2(x)$

$=> \cos^2(x) = \dfrac{1+\cos(2x)}{2}$

$\mathbf{\sin^2(x) = \frac{1-\cos(2x)}{2}}$ **(Identity 23)**

From

$\cos(2x) = 1 - 2.\sin^2(x)$

=> $\cos(2x) - 1 = -2.\sin^2(x)$

=> $-\cos(2x) + 1 = 2.\sin^2(x)$ (Multiplied both sides by -1)

=> $\dfrac{-\cos(2x) + 1}{2} = \sin^2(x)$

=> $\sin^2(x) = \frac{1-\cos(2x)}{2}$ (Rearrange)

$\boldsymbol{sin(x).\cos(x) = \frac{1}{2}(\sin(x + y) + \sin(x - y))}$ **(Identity 24)**

We have already established the following:

$\sin(x + y) = \sin(x).\cos(y) + \sin(y).\cos(x)$

$\sin(x - y) = \sin(x).\cos(y) - \sin(y).\cos(x)$

Now let's add the left-hand sides of the aforementioned equations while adding the right-hand side of them together to get the following:

$\sin(x + y) + \sin(x - y) = \sin(x).\cos(y) + \sin(y).\cos(x) + \sin(x).\cos(y) - \sin(y).\cos(x)$

$\sin(x + y) + \sin(x - y) = 2.\sin(x).\cos(y)$

$\sin(x).\cos(y) = \dfrac{1}{2}(\sin(x + y) + \sin(x - y))$

$cos(x).\cos(y) = \frac{1}{2}(\cos(x+y) + \cos(x-y))$ **(Identity 25)**

We have already established the following:

$$\cos(x+y) = \cos(x).\cos(y) - \sin(x).\sin(y)$$

$$\cos(x-y) = \cos(x).\cos(y) + \sin(x).\sin(y)$$

Now let's add the left-hand sides of the aforementioned equations while adding the right-hand side of them together, and we get the following:

$$\cos(x+y) + \cos(x-y)$$
$$= \cos(x).\cos(y) - \sin(x).\sin(y) + \cos(x).\cos(y) + \sin(x).\sin(y)$$

$$\cos(x+y) + \cos(x-y) = 2.\cos(x).\cos(y)$$

$$=> \cos(x).\cos(y) = \frac{1}{2}(\cos(x+y) + \cos(x-y))$$

$sin(x).\sin(y) = \frac{1}{2}(\cos(x-y) - \cos(x+y))$ **(Identity 26)**

We have already established the following:

$$\cos(x+y) = \cos(x).\cos(y) - \sin(x).\sin(y)$$

$$\cos(x-y) = \cos(x).\cos(y) + \sin(x).\sin(y)$$

Now let's subtract the left-hand sides of the aforementioned equations while subtracting the right-hand side of them together, and we get the following:

$$\cos(x+y) - \cos(x-y)$$

$$= \cos(x).\cos(y) \ - \ \sin(x).\sin(y) - \ \cos(x).\cos(y) - \sin(x).\sin(y)$$

$$\cos(x+y) - \cos(x+y) \ = \ -2.\sin(x).\sin(y)$$

Switch sides

$$2.\sin(x).\sin(y) = -\cos(x+y) + \cos(x-y)$$

$$=> 2.\sin(x).\sin(y) = \cos(x-y) \ - \cos(x+y)$$

Divide both sides by 2

$$=> \sin(x).\sin(y) = \frac{1}{2}(\cos(x-y) \ - \cos(x+y))$$

3.7 Identity Exercises:

Prove the following identities:

1. $\mathbf{tan^2(x) - sin^2(x) = tan^2(x) . sin^2(x)}$

Solution:

Starting from the left side:

$$\tan^2(x) - \sin^2(x) = \frac{\sin^2(x)}{\cos^2(x)} - \sin^2(x) =$$

Getting common denominator to be $\cos^2(x)$:

$$\frac{\sin^2(x)}{\cos^2(x)} - \frac{\cos^2(x) . \sin^2(x)}{\cos^2(x)} = \frac{\sin^2(x) - \cos^2(x) . \sin^2(x)}{\cos^2(x)}$$

Factor out $\sin^2(x)$

$$\frac{\sin^2(x)\,(1 - \cos^2(x))}{\cos^2(x)}$$

Note $\sin^2(x) = 1 - \cos^2(x)$

$$\frac{\sin^2(x)\sin^2(x)}{\cos^2(x)} = \frac{\sin^2(x)}{\cos^2(x)}\sin^2(x) = \tan^2(x) . \sin^2(x)$$

---

2. $\mathbf{cot(2x)} = \frac{\mathbf{1 - tan^2(x)}}{\mathbf{2tan(x)}}$

Solution:

$$\cot(2x) = \frac{\cos(2x)}{\sin(2x)} = \frac{\cos^2(x) - \sin^2(x)}{2\sin(x)\cos(x)}$$

Divide the denominator and numerator by $\cos^2(x)$

$$\frac{\frac{\cos^2(x)}{\cos^2(x)} - \frac{\sin^2(x)}{\cos^2(x)}}{\frac{2\sin(x)\cos(x)}{\cos^2(x)}} = \frac{1 - \tan^2(x)}{\frac{2\sin(x)}{\cos(x)}} = \frac{1 - \tan^2(x)}{2\tan(x)}$$

---

3. $\mathbf{\csc(2x)} = \frac{\mathbf{sec(x)}}{\mathbf{2sin(x)}}$

Solution:

$$\csc(2x) = \frac{1}{\sin(2x)} = \frac{1}{2\sin(x).\cos(x)} = \frac{1}{2\sin(x)}.\frac{1}{\cos(x)} = \frac{1}{2\sin(x)}.\sec(x)$$

$$= \frac{\sec(x)}{2\sin(x)}$$

---

4. $\frac{\sin(3x)+\sin(7x)}{\cos(3x)-\cos(7x)} = \cot(2x)$

Solution:

If you remember, it is great; if not (like me), let's build it :D (look at identity 24 and 26 to see how

$$sin(a).\cos(b) = \frac{1}{2}(\sin(a+b) + \sin(a-b))$$

$$\sin(a).\sin(b) = \frac{1}{2}(\cos(a-b) - \cos(a+b))$$

Note we are looking for two numbers whose addition is 7 and their difference is 3

$$\begin{cases} a + b = 7x \\ a - b = 3x \end{cases}$$

From the second one, we get:

$$a - b = 3x => a = b + 3x$$

We substitute this in the second one, and we get:

$$a + b = 7x => b + 3x + b = 7x => 2b = 7x - 3x => 2b = 4x => b = \frac{4x}{2} = 2x$$

$$a = b + 3x => a = 2x + 3x = 5x$$

Therefore, we can substitute "a" and "b" as follows

$$sin(5x).\cos(2x) = \frac{1}{2}(\sin(5x+2x) + \sin(5x-3x)):$$

$$=> sin(5x).\cos(2x) = \frac{1}{2}(\sin(7x) + \sin(4x))$$

$$=> 2sin(5x).\cos(2x) = \sin(7x) + \sin(4x)$$

$$\sin(5x).\sin(2x) = \frac{1}{2}(\cos(5x-2x) - \cos(5x+2x))$$

$$=> \sin(5x).\sin(2x) = \frac{1}{2}(\cos(3x) - \cos(7x))$$

$$=> 2sin(5x).\sin(2x) = \cos(3x) - \cos(7x)$$

Getting back to the original question:

$$\frac{\sin(3x) + \sin(7x)}{\cos(3x) - \cos(7x)} = \frac{2sin(5x).\cos(2x)}{2sin(5x).\sin(2x)}$$

$$= \frac{\cos(2x)}{\sin(2x)} = \cot(2x)$$

---

5. $\frac{\csc(x)+\cot(x)}{\tan(x)+\sin(x)} = \mathbf{\cot(x)\,.\,\csc(x)}$

Solution:

$$= \frac{\csc(x) + \cot(x)}{\tan(x) + \sin(x)} = \frac{\frac{1}{\sin(x)} + \frac{\cos(x)}{\sin(x)}}{\frac{\sin(x)}{\cos(x)} + \sin(x)}$$

Now cos(x) is the common denominator in the bottom

$$= \frac{\frac{1 + \cos(x)}{\sin(x)}}{\frac{\sin(x) + \sin(x)\cos(x)}{\cos(x)}}$$

Factor out sin(x) in the denominator

$$= \frac{\frac{1 + \cos(x)}{\sin(x)}}{\frac{\sin(x)\,(1 + \cos(x))}{\cos(x)}}$$

Turn division into multiplication by multiplying the numerator by the reciprocal of the denominator.

$$= \frac{(1 + \cos(x))}{\sin(x)} . \frac{\cos(x)}{\sin(x)\,(1 + \cos(x))}$$

Note you can cancel $1 + \cos(x)$

$$= \frac{1}{\sin(x)} \cdot \frac{\cos(x)}{\sin(x)} = \csc(x) \,.\, \cot(x) = \cot(x) \,.\, \csc(x)$$

---

6. $\mathbf{cot}(\boldsymbol{x}) + \mathbf{tan}(\boldsymbol{x}) = \mathbf{sec}(\boldsymbol{x}) \,.\, \mathbf{csc}(\boldsymbol{x})$

Solution:

$$\cot(x) + \tan(x) = \frac{\cos(x)}{\sin(x)} + \frac{\sin(x)}{\cos(x)}$$

Getting the common denominator, which is sin(x) cos(x):

$$= \frac{\cos^2(x) + \sin^2(x)}{\sin(x) \cos(x)} = \frac{1}{\sin(x) \,.\, \cos(x)} = \frac{1}{\sin(x)} \cdot \frac{1}{\cos(x)} = \csc(x) \,.\, \sec(x) = \sec(x) \,.\, \csc(x)$$

---

7. $\mathbf{cos}\left(\frac{\pi}{2} - \boldsymbol{x}\right) = \mathbf{sin}(\boldsymbol{x})$

Solution:

$$\cos\left(\frac{\pi}{2} - x\right) = \cos\left(\frac{\pi}{2}\right) . \cos(x) + \sin\left(\frac{\pi}{2}\right) . \sin(x)$$

Since $\cos\left(\frac{\pi}{2}\right) = 0$ and $\sin\left(\frac{\pi}{2}\right) = 1$

$$= 0. \cos(x) + 1. \sin(x) = \sin(x)$$

---

8. $\mathbf{\sin\left(\frac{\pi}{2}+x\right)=\cos(x)}$

Solution:

$$\sin\left(\frac{\pi}{2}+x\right)=sin\left(\frac{\pi}{2}\right)\cos(x)+\sin(x).\cos\left(\frac{\pi}{2}\right)$$

Since $\cos\left(\frac{\pi}{2}\right)=0$ and $\sin\left(\frac{\pi}{2}\right)=1$

$$1.\cos(x)+\sin(x).0=\cos(x)$$

---

9. $\mathbf{\cos(x)\cot(x)=\cos(x)}$

Solution:

$$\sin(x)\cot(x)=\sin(x).\frac{\cos(x)}{\sin(x)}=\cos(x)$$

---

10. $\mathbf{\sec(x)-\cos(x)=\tan(x).\sin(x)}$

Solutions:

$$\sec(x)-\cos(x)=\frac{1}{\cos(x)}-\cos(x)$$

The common denominator would be cos(x).

$$\frac{1}{\cos(x)} - \frac{\cos(x)}{\cos(x)} = \frac{1-\cos^2(x)}{\cos(x)}$$

From rearranging the first identity, $1 - \cos^2(x) = \sin^2(x)$

$$\frac{\sin^2(x)}{\cos(x)} = \frac{\sin(\mathrm{x})}{\cos(x)}.\sin(x) = \tan(x).\sin(x)$$

---

11. $\mathbf{cot^2(x) + sec^2(x) = tan^2(x) + csc^2(x)}$

Solution:

$$\cot^2(x) + \sec^2(x) = \frac{\cos^2(x)}{\sin^2(x)} + \frac{1}{\cos^2(x)}$$

From rearranging the first identity, $1 - \sin^2(x) = \cos^2(x)$

$$= \frac{1-\sin^2(x)}{\sin^2(x)} + \frac{1}{\cos^2(x)} = \frac{1}{\sin^2(x)} - \frac{\sin^2(x)}{\sin^2(x)} + \frac{1}{\cos^2(x)}$$

$$= \csc^2(x) - 1 + \frac{1}{\cos^2(x)}$$

Getting the common denominator between the last two terms, it leads to $\cos^2(x)$

$$= \csc^2(x) + \frac{-\cos^2(x)}{\cos^2(x)} + \frac{1}{\cos^2(x)} = \csc^2(x) + \frac{-\cos^2(x)+1}{\cos^2(x)}$$

From rearranging the first identity, $1 - \cos^2(x) = \sin^2(x)$

$$= \csc^2(x) + \frac{\sin^2(x)}{\cos^2(x)} = \csc^2(x) + \tan^2(x) = \tan^2(x) + \csc^2(x)$$

---

12. $\mathbf{2\csc(2x) = sec(x).csc(x)}$

Solution:

$$2\csc(2x) = \frac{2}{\sin(2x)} = \frac{2}{2\sin(x)\cos(x)} = \frac{1}{\sin(x)\cos(x)}$$

$$= \frac{1}{\sin(x)} \cdot \frac{1}{\cos(x)} = \csc(x).\sec(x) = \sec(x).\csc(x)$$

---

13. $\mathbf{\tan(2x)} = \frac{\mathbf{2.tan(x)}}{\mathbf{1-\tan^2(x)}}$

Solution:

$$\tan(2x) = \frac{\sin(2x)}{\cos(2x)} = \frac{2\sin(x)\cos(x)}{\cos^2(x) - \sin^2(x)}$$

Divide the numerator and denominator by $\cos^2(x)$

$$= \frac{\frac{2\sin(x)\cos(x)}{\cos^2(x)}}{\frac{\cos^2(x)}{\cos^2(x)} - \frac{\sin^2(x)}{\cos^2(x)}} = \frac{\frac{2\sin(x)}{\cos(x)}}{1 - \tan^2(x)} = \frac{2\tan(x)}{1 - \tan^2(x)}$$

---

14. $\frac{\mathbf{1}}{\mathbf{1-sin(x)}} + \frac{\mathbf{1}}{\mathbf{1+sin(x)}} = \mathbf{2.\sec^2(x)}$

Solution:

$$\frac{1}{1 - sin(x)} + \frac{1}{1 + sin(x)}$$

The common denominator would be $(1 - sin(x)).(1 + sin(x))$

$$\frac{1}{1-sin(x)}+\frac{1}{1+sin(x)}=\frac{1+sin(x)+1-sin(x)}{(1-sin(x))(1+sin(x))}=$$

Also we know $(1-sin(x)).(1+sin(x)) = 1^2 - \sin^2(x) = \cos^2(x)$

$$\frac{2}{\cos^2(x)} = 2.\sec^2(x)$$

---

15. $\mathbf{\sin(x).\sin(2x)+\cos(x).\cos(2x)=\cos(x)}$

Solution:

$\sin(x).\sin(2x)+\cos(x).\cos(2x)$

$= \sin(x)\,(2\sin(x)\cos(x)) + \cos(x)\,(\cos^2(x)-\sin^2(x))$

$= 2\sin^2(x)\cos(x) + \cos(x)\,(\cos^2(x)-\sin^2(x))$

Factor out cos(x)

$= \cos(x)\,(2\sin^2(x)+\cos^2(x)-\sin^2(x))$

$= \cos(x)\,(\sin^2(x)+\cos^2(x))$

From the first identity, we have $\sin^2(x)+\cos^2(x)=1$

$= \cos(x)\,(1) = \cos(x)$

---

16. $\mathbf{\sin^2(x)-\sin^2(y)=\sin(x+y).\sin(x-y)}$

Solution:

Note that we can start from the right side and get to the left side.

$\sin(x+y).\sin(x-y)$

$= (\sin(x)\cos(y) + \sin(y)\cos(x)).(\sin(x)\cos(y) - \sin(y)\cos(x))$

Just multiply each term (you should get four terms and then simplify). You also can use the following as a reference to speed up the process of multiplication ($(a+b)(a-b) = a^2 - b^2$

$= \sin^2(x)\cos^2(y) - \sin^2(y)\cos^2(x) =$

From rearranging the first identity,

$1 - \sin^2(y) = \cos^2(y)$ and $1 - \sin^2(x) = \cos^2(x)$

We can conclude:

$= \sin^2(x)\,(1 - \sin^2(y)) - \sin^2(y)\,(1 - \sin^2(x))$

$= \sin^2(x) - \sin^2(x)\sin^2(y) - \sin^2(y) + \sin^2(y)\sin^2(x)$

$= \sin^2(x) - \sin^2(y)$

---

17. $\frac{\sin(x)}{1-\cos(x)} = \csc(x) + \cot(x)$

Solution:

$$\frac{\sin(x)}{1-\cos(x)} =$$

Multiply the numerator and denominator by $1 + \cos(x)$

$$\frac{\sin(x)}{(1-\cos(x))}\frac{(1+\cos(x))}{(1+\cos(x))} = \frac{\sin(x)(1+\cos(x))}{(1-\cos^2(x))}$$

From rearranging the first identity, $1 - \cos^2(x) = \sin^2(x)$

$$= \frac{\sin(x) + \sin(x)\cos(x))}{\sin^2(x)} = \frac{\sin(x)}{\sin^2(x)} + \frac{\sin(x)\cos(x))}{\sin^2(x)}$$

$$= \frac{1}{\sin(x)} + \frac{\cos(x)}{\sin(x)} = \csc(x) + \cot(x)$$

---

18. $\mathbf{\tan(x) + \tan(y) = \frac{\sin(x+y)}{\cos(x).\cos(y)}}$

Solution:

Let's start from the right side:

$$\frac{\sin(x+y)}{\cos(x).\cos(y)} = \frac{\sin(\mathrm{x}).\cos(\mathrm{y}) + \sin(\mathrm{y})\cos(x)}{\cos(x).\cos(y)}$$

$$= \frac{\sin(\mathrm{x}).\cos(\mathrm{y})}{\cos(x).\cos(y)} + \frac{\sin(\mathrm{y})\cos(x)}{\cos(x).\cos(y)} = \frac{\sin(\mathrm{x})}{\cos(x)} + \frac{\sin(\mathrm{y})}{\cos(y)}$$

$$= \tan(x) + \tan(y)$$

---

19. $\mathbf{\sin(3x) + \sin(x) = 2.\sin(2x).\cos(x)}$

Solution:

$$\sin(3x) + \sin(x) = \sin(2x + x) + \sin(x)$$

$$= \sin(2x).\cos(x) + \sin(\mathrm{x}).\cos(2x) + \sin(x)$$

Factor out sin(x) from the last two terms:

$$= \sin(2x).\cos(x) + \sin(\mathrm{x})\,(\cos(2x) + 1)$$

We use $\cos(2x) = \cos^2(x) - \sin^2(x)$ and $1 = \cos^2(x) + \sin^2(x)$ in the second term to achieve the following:

$$= \sin(2x).\cos(x) + \sin(\mathrm{x})\,(\cos^2(x) - \sin^2(x) + \cos^2(x) + \sin^2(x))$$

$$= \sin(2x).\cos(x) + \sin(\mathrm{x})\,(\cos^2(x) + \cos^2(x))$$

$$= \sin(2x).\cos(x) + \sin(\mathrm{x})\,(2.\cos^2(x))$$

Factor out cos(x)

$$= \cos(x)\,(\sin(2x) + 2.\sin(\mathrm{x})\cos(x)) = \cos(x)\,(\sin(2x) + \sin(2\mathrm{x}))$$

$$= \cos(x)\,(2\sin(2x)) = 2.\sin(2x).\cos(x)$$

---

20. $\mathbf{\cos(3x) = 4.\cos^3(x) - 3\cos(x)}$

Solution:

$$\cos(3x) = \cos(2x + x) = \cos(2x)\cos(x) - \sin(2x)\sin(x)$$

$$= (\cos(x)\cos(x) - \sin(x)\sin(x))\cos(x) - (2\sin(x)\cos(x)\sin(x))$$

$$= (\cos^2(x) - \sin^2(x))\cos(x) - (2\sin^2(x)\cos(x))$$

$$= \cos^3(x) - \sin^2(x)\cos(x) - 2\sin^2(x)\cos(x)$$

$$= \cos^3(x) - 3\sin^2(x)\cos(x)$$

From rearranging the first identity, $1 - \cos^2(x) = \sin^2(x)$

$$= \cos^3(x) + (1 - \cos^2(x))(-3\cos(x))$$

$$= \cos^3(x) - 3\cos(x) + 3\cos^3(x) = 4.\cos^3(x) - 3\cos(x)$$

---

21. $\frac{1+\tan(x)}{1+\cot(x)} = \frac{1-\tan(x)}{\cot(x)-1}$

Solution:

This time, let's simplify both sides starting from the left-hand side.

$$\frac{1+\tan(x)}{1+\cot(x)} = \frac{1+\frac{\sin(x)}{\cos(x)}}{1+\frac{\cos(x)}{\sin(x)}}$$

The common denominator of the numerator is cos(x), and the common denominator of the denominator is sin(x).

$$= \frac{\frac{\cos(x)+\sin(x)}{\cos(x)}}{\frac{\sin(x)+\cos(x)}{\sin(x)}}$$

Turn division into multiplication by multiplying the numerator by the reciprocal of denominator

$$= \frac{\cos(x)+\sin(x)}{\cos(x)} \cdot \frac{\sin(x)}{\sin(x)+\cos(x)} = \frac{\sin(x)}{\cos(x)} = \tan(x)$$

Now let's see if we get the same value for the right-hand side:

$$\frac{1-\tan(x)}{\cot(x)-1} = \frac{1-\frac{\sin(x)}{\cos(x)}}{\frac{\cos(x)}{\sin(x)}-1}$$

The common denominator of the numerator is cos(x) and the common denominator of the denominator sin(x).

$$= \frac{\frac{\cos(x) - \sin(x)}{\cos(x)}}{\frac{\cos(x) - \sin(x)}{\sin(x)}}$$

Turn division into multiplication by multiplying the numerator by the reciprocal of the denominator

$$= \frac{\cos(x) - \sin(x)}{\cos(x)} \cdot \frac{\sin(x)}{\cos(x) - \sin(x)} = \frac{\sin(x)}{\cos(x)} = \tan(x)$$

Therefore, the left-hand side is equal to the right-hand side:

---

22. $\frac{\cos^2(x) - \sin^2(x)}{\cos^2(x) + \sin(x).\cos(x)} = 1 - \tan(x)$

Solution:

$$\frac{\cos^2(x) - \sin^2(x)}{\cos^2(x) + \sin(x).\cos(x)}$$

Note $\cos^2(x) - \sin^2(x)$ is similar to $a^2 - b^2 = (a - b)(a + b)$. That is, $\cos^2(x) - \sin^2(x) = (\cos(x) - sin\ (x))(\cos(x) + sin\ (x))$

$$= \frac{(\cos(x) - sin\ (x))(\cos(x) + sin\ (x))}{(cos(x) + \sin(x)).\cos(x)}$$

Cancel out $(cos(x) + \sin(x))$ from the top and the bottom to achieve the following:

$$= \frac{\cos(x) - sin\ (x)}{\cos(x)} = \frac{\cos(x)}{\cos(x)} - \frac{sin\ (x)}{\cos(x)} = 1 - \tan(x)$$

---

23. $\mathbf{\tan^2(x) - \cos^2(x) = \frac{1}{\cos^2(x)} - 1 - \cos^2(x)}$

Solution:

$$\tan^2(x) - \cos^2(x) = \frac{\sin^2(x)}{\cos^2(x)} - \cos^2(x)$$

From rearranging the first identity, $1 - \cos^2(x) = \sin^2(x)$

$$\frac{1 - \cos^2(x)}{\cos^2(x)} - \cos^2(x) = \frac{1}{\cos^2(x)} - 1 - \cos^2(x)$$

---

24. $\mathbf{\frac{1}{1+\cos(x)} + \frac{1}{1-\cos(x)} = \frac{2}{\sin^2(x)}}$

Solution:

$$\frac{1}{1 + \cos(x)} + \frac{1}{1 - \cos(x)}$$

The common denominator would be $(1 + \cos(x))(1 - \cos(x))$, which, after multiplication, becomes $1^2 - \cos^2(x)$. You can use the difference of the square to find the answer $(a + b)(a - b) = a^2 - b^2$

From rearranging the first identity, we achieve the following:

$$1^2 - \cos^2(x) = \sin^2(x)$$

$$\frac{1}{1 + \cos(x)} + \frac{1}{1 - \cos(x)}$$

$$= \frac{1-\cos(\mathrm{x})}{(1+\cos(\mathrm{x}))(1-\cos(\mathrm{x}))} + \frac{1+\cos(\mathrm{x})}{(1+\cos(\mathrm{x}))(1-\cos(\mathrm{x}))}$$

$$= \frac{1-\cos(\mathrm{x})}{\sin^2(x)} + \frac{1+\cos(\mathrm{x})}{\sin^2(x)} = \frac{2}{\sin^2(x)}$$

---

25. $\boldsymbol{cos(x).\tan^3(x) = \sin(\mathrm{x})\tan^2(x)}$

Solution:

$cos(x).\tan^3(x) = cos(x).\tan(x).\tan^2(x)$

$$= cos(x).\frac{\sin(x)}{\cos(x)}.\tan^2(x) = \sin(x).\tan^2(x)$$

---

26. $\boldsymbol{\sin^2(x) + \cos^4(x) = \cos^2(x) + \sin^4(x)}$

Solution:

$\sin^2(x) + \cos^4(x) = \sin^2(x) + (\cos^2(x))^2$

From rearranging the first identity, $1 - \sin^2(x) = \cos^2(\mathrm{x})$

$\sin^2(x) + (1 - \sin^2(x))^2 = \sin^2(x) + 1 + \sin^4(x) - 2\sin^2(x)$

$= 1 - \sin^2(x) + \sin^4(x)$

From rearranging the first identity, $1 - \sin^2(x) = \cos^2(\mathrm{x})$

$= \cos^2(\mathrm{x}) + \sin^4(x)$

---

27. $\mathbf{(\sin(x)+\cos(x))}\left(\frac{\tan^2(x)+1}{\tan(x)}\right)=\frac{1}{\cos(x)}+\frac{1}{\sin(x)}$

Solution:

$$(\sin(x)+\cos(x))\left(\frac{tan^2(x)+1}{tan(x)}\right)$$

$$=(\sin(x)+\cos(x))\left(\frac{tan^2(x)}{tan(x)}+\frac{1}{tan(x)}\right)$$

$$=(\sin(x)+\cos(x))\left(tan(x)+\frac{1}{tan(x)}\right)$$

$$=(\sin(x)+\cos(x))\left(\frac{\sin(x)}{\cos(x)}+\frac{1}{\frac{\sin(x)}{\cos(x)}}\right)$$

Note we rewrite the second term in the second bracket as $\frac{\cos(x)}{\sin(x)}$. Then we proceed to get the common denominator in the second bracket.

$$=(\sin(x)+\cos(x))\left(\frac{\sin^2(x)}{\cos(x).\sin(x)}+\frac{\cos^2(x)}{\cos(x).\sin(x)}\right)$$

$$=(\sin(x)+\cos(x))\left(\frac{\sin^2(x)+\cos^2(x)}{\cos(x).\sin(x)}\right)$$

From the first identity, we have $\sin^2(x)+\cos^2(x)=1$

$$=(\sin(x)+\cos(x))\left(\frac{1}{\cos(x).\sin(x)}\right)$$

$$=\frac{\sin(x)}{\cos(x).\sin(x)}+\frac{\cos(x)}{\cos(x).\sin(x)}=\frac{1}{\cos(x)}+\frac{1}{\sin(x)}$$

---

28. $\mathbf{\tan^2(x) + \cos^2(x) + \sin^2(x)} = \frac{\mathbf{1}}{\mathbf{\cos^2(x)}}$

Solution:

$\tan^2(x) + \cos^2(x) + \sin^2(x)$

From the first identity, we have $\cos^2(x) + \sin^2(x) = 1$

$$= \tan^2(x) + 1 = \frac{\sin^2(x)}{\cos^2(x)} + 1$$

Rearranging the first identity, we achieve the following:

$\sin^2(x) = 1 - \cos^2(x)$

$$= \frac{1 - \cos^2(x)}{\cos^2(x)} + 1 = \frac{1}{\cos^2(x)} - \frac{\cos^2(x)}{\cos^2(x)} + 1 = \frac{1}{\cos^2(x)} - 1 + 1$$

$$= \frac{1}{\cos^2(x)}$$

---

29. $\mathbf{\sin\left(\frac{\pi}{4} + x\right) + \cos\left(\frac{\pi}{4} - x\right) = \sqrt{2}\cos(x)}$

Solution:

$$\sin\left(\frac{\pi}{4} + x\right) + \cos\left(\frac{\pi}{4} - x\right)$$

$$= \sin\left(\frac{\pi}{4}\right).\cos(x) - \sin(x).\cos\left(\frac{\pi}{4}\right) + \cos\left(\frac{\pi}{4}\right)\cos(x) + \sin\left(\frac{\pi}{4}\right)\sin(x)$$

We have $\sin\left(\frac{\pi}{4}\right) = \cos\left(\frac{\pi}{4}\right) = \frac{\sqrt{2}}{2}$

$$= \frac{\sqrt{2}}{2}.\cos(x) - \frac{\sqrt{2}}{2}\sin(x) + \frac{\sqrt{2}}{2}\cos(x) + \frac{\sqrt{2}}{2}\sin(x)$$

$$= \frac{\sqrt{2}}{2}.\cos(x) + \frac{\sqrt{2}}{2}.\cos(x) = 2\frac{\sqrt{2}}{2}.\cos(x) = \sqrt{2}\cos(x)$$

---

30. $\mathbf{\sin\left(\frac{\pi}{2} - x\right).\cot\left(\frac{\pi}{2} + x\right) = -\sin(x)}$

Solution:

$$\sin\left(\frac{\pi}{2} - x\right) + \cot\left(\frac{\pi}{2} + x\right)$$

$$= \left(\sin\left(\frac{\pi}{2}\right).\cos(x) - \sin(x).\cos\left(\frac{\pi}{2}\right)\right).\left(\frac{\cos\left(\frac{\pi}{2}\right)\cos(x) - \sin\left(\frac{\pi}{2}\right)\sin(x)}{\sin\left(\frac{\pi}{2}\right).\cos(x) + \sin(x).\cos\left(\frac{\pi}{2}\right)}\right)$$

We have $\sin\left(\frac{\pi}{2}\right) = 1, and \cos\left(\frac{\pi}{2}\right) = 0$

$$= (1.\cos(x) - 0.\sin(x)).\left(\frac{0\cos(x) - 1.\sin(x)}{1.\cos(x) + \sin(x).0}\right)$$

$$= \cos(x)\left(-\frac{\sin(x)}{\cos(x)}\right) = -\sin(x)$$

---

31. $\mathbf{\frac{\cos(2x)+1}{\sin(2x)} = \cot(x)}$

Solution:

$$\frac{\cos(2x) + 1}{\sin(2x)} = \frac{\cos^2(x) - \sin^2(x) + 1}{2\cos(x).\sin(x)}$$

Note by rearranging identity one, we achieve:

$-\sin^2(x) + 1 = \cos^2(x)$

$$\frac{\cos^2(x) + \cos^2(x)}{2\cos(x).\sin(x)} = \frac{2\cos^2(x)}{2\cos(x).\sin(x)} = \frac{\cos(x)}{\sin(x)} = \cot(x)$$

---

32. $\frac{\sin(2x)}{1-\cos(2x)} = \cot(x)$

Solution:

$$\frac{\sin(2x)}{1-\cos(2x)} = \frac{2\sin(x).\cos(x)}{1-\cos^2(x)+\sin^2(x)} =$$

From rearranging the first identity: $1 - \cos^2(x) = \sin^2(x)$

$$= \frac{2\sin(x).\cos(x)}{\sin^2(x)+\sin^2(x)} = \frac{2\sin(x).\cos(x)}{2\sin^2(x)} = \frac{\cos(x)}{\sin(x)} = \cot(x)$$

---

33. $(\sin(x) + \cos(x))^2 = 1 + \sin(2x)$

Solution:

$(\sin(x) + \cos(x))^2 = \sin^2(x) + \cos^2(x) + 2\sin(x).\cos(x)$

From $2\sin(x).\cos(x) = \sin(2x)$

$1 + 2\sin(x).\cos(x) = 1 + \sin(2x)$

---

34. $\cos^4(x) - \sin^4(x) = \cos(2x)$

Solution:

$\cos^4(x) - \sin^4(x) = (\cos^2(x) - \sin^2(x))(\cos^2(x) + \sin^2(x)) =$

Since $(\cos^2(x) - \sin^2(x)) = \cos(2x)$ and $(\cos^2(x) + \sin^2(x)) = 1$

$\cos(2x)(1) = \cos(2x)$

---

35. $\mathbf{cot(x) - tan(\mathit{x}) = 2.cot(2\mathit{x})}$

Solution:

$$\cot(x) - \tan(x) = \frac{\cos(x)}{\sin(x)} - \frac{\sin(x)}{\cos(x)}$$

Choosing the common denominator to be $\sin(x).\cos(x)$

$$= \frac{\cos^2(x)}{\sin(x).\cos(x)} - \frac{\sin^2(x)}{\sin(x).\cos(x)}$$

$$= \frac{\cos^2(x) - \sin^2(x)}{\sin(x).\cos(x)}$$

Note $\cos^2(x) - \sin^2(x) = \cos(2x)$, and $\sin(x).\cos(x) = \frac{1}{2}\sin(2x)$

$$= \frac{\cos(2x)}{\frac{1}{2}\sin(2x)} = 2\frac{\cos(2x)}{\sin(2x)} \text{ (Note } \frac{1}{\frac{1}{2}} = 2)$$

$$= 2\cot(2x)$$

---

36. $\mathbf{cot(x) + tan(x) = 2.csc(2x)}$

Solution:

$$\cot(x) + \tan(x) = \frac{\cos(x)}{\sin(x)} + \frac{\sin(x)}{\cos(x)}$$

The common denominator is $\sin(x).\cos(x)$

$$= \frac{\cos^2(x)}{\sin(x).\cos(x)} + \frac{\sin^2(x)}{\sin(x).\cos(x)} = \frac{\cos^2(x) + \sin^2(x)}{\sin(x).\cos(x)} = \frac{1}{\sin(x).\cos(x)}$$

Note that $\sin(2x) = 2\sin(x).\cos(x)$, divide both sides by 2

$$\frac{1}{2}\sin(2x) = \sin(x).\cos(x)$$

$$=> \frac{1}{\sin(x).\cos(x)} = \frac{1}{\frac{1}{2}\sin(2x)}$$

Note $\frac{1}{\frac{1}{2}} = 2$

$$= \frac{2}{\sin(2x)} = 2\csc(2x)$$

---

37. $\mathbf{\frac{1+tan(x)}{1-tan(x)} = tan\left(x + \frac{\pi}{4}\right)}$

Solution:

Let's start from the right-hand side:

$$\tan\left(x+\frac{\pi}{4}\right) = \frac{\sin\left(x+\frac{\pi}{4}\right)}{\cos\left(x+\frac{\pi}{4}\right)} = \frac{\sin(x).\cos\left(\frac{\pi}{4}\right)+\sin\left(\frac{\pi}{4}\right).\cos(x)}{\cos(x).\cos\left(\frac{\pi}{4}\right)-\sin(x).\sin\left(\frac{\pi}{4}\right)}$$

Note $\sin\left(\frac{\pi}{4}\right) = \cos\left(\frac{\pi}{4}\right) = \frac{\sqrt{2}}{2}$

$$= \frac{\sin(x).\frac{\sqrt{2}}{2}+\frac{\sqrt{2}}{2}.\cos(x)}{\cos(x).\frac{\sqrt{2}}{2}-\sin(x).\frac{\sqrt{2}}{2}}$$

Divide the top and bottom by $\frac{\sqrt{2}}{2}$

$$= \frac{\frac{\frac{\sqrt{2}}{2}\sin(x)}{\frac{\sqrt{2}}{2}}+\frac{\frac{\sqrt{2}}{2}\cos(x)}{\frac{\sqrt{2}}{2}}}{\frac{\frac{\sqrt{2}}{2}\cos(x)}{\frac{\sqrt{2}}{2}}-\frac{\frac{\sqrt{2}}{2}\sin(x)}{\frac{\sqrt{2}}{2}}} = \frac{\sin(x)+\cos(x)}{\cos(x)-\sin(x)}$$

Divide the top and bottom by $\cos(x)$

$$= \frac{\frac{\sin(x)}{\cos(x)}+\frac{\cos(x)}{\cos(x)}}{\frac{\cos(x)}{\cos(x)}-\frac{\sin(x)}{\cos(x)}} = \frac{\tan(x)+1}{1-\tan(x)} = \frac{1+\tan(x)}{1-\tan(x)}$$

---

38. $\mathbf{\csc(2x)+\cot(2x)=\cot(x)}$

Solution:

$$\csc(2x)+\cot(2x) = \frac{1}{\sin(2x)}+\frac{\cos(2x)}{\sin(2x)} = \frac{1+\cos(2x)}{\sin(2x)}$$

$$= \frac{1 + \cos^2(x) - \sin^2(x)}{2\sin(x).\cos(x)}$$

Rearranging the first identity: $1 - \sin^2(x) = \cos^2(x)$

$$= \frac{\cos^2(x) + \cos^2(x)}{2\sin(x).\cos(x)} = \frac{2\cos^2(x)}{2\sin(x).\cos(x)} = \frac{cos(x)}{\sin(x)}$$

$$= \cot(x)$$

---

39. $\frac{\mathbf{2tan}(x)}{\mathbf{1+tan^2}(x)} = \mathbf{sin(2x)}$

Solution:

$$\frac{2\tan(x)}{1 + \tan^2(x)} = \frac{\frac{2\sin(x)}{\cos(x)}}{1 + \frac{\sin^2(x)}{\cos^2(x)}} =$$

Choosing the common denominator to be $\cos^2(x)$ for the bottom part.

$$\frac{\frac{2\sin(x)}{\cos(x)}}{\frac{\cos^2(x)}{\cos^2(x)} + \frac{\sin^2(x)}{\cos^2(x)}} = \frac{\frac{2\sin(x)}{\cos(x)}}{\frac{\cos^2(x) + \sin^2(x)}{\cos^2(x)}} = \frac{\frac{2\sin(x)}{\cos(x)}}{\frac{1}{\cos^2(x)}}$$

Turning division into multiplication (multiplying the numerator by the reciprocal of the denominator)

$$\frac{2\sin(x)}{\cos(x)} \cdot \frac{\cos^2(x)}{1} = 2\sin(x)\cos(x) = \sin(2x)$$

---

40. $\mathbf{sec(2}x\mathbf{)} = \frac{\mathbf{csc}(x)}{\mathbf{csc}(x)\mathbf{-2.sin}(x)}$

Solution:

Starting from the right-hand side:

$$\frac{\csc(x)}{\csc(x) - 2.\sin(x)} = \frac{\frac{1}{\sin(x)}}{\frac{1}{\sin(x)} - 2\sin(x)}$$

Choose sin(x) as the common denominator in the bottom of the fraction:

$$= \frac{\frac{1}{\sin(x)}}{\frac{1}{\sin(x)} - \frac{2\sin^2(x)}{\sin(x)}} = \frac{\frac{1}{\sin(x)}}{\frac{1-2\sin^2(x)}{\sin(x)}}$$

Replacing 1 in the denominator with $\sin^2(x) + \cos^2(x) = 1$ based on the first identity.

$$\frac{\frac{1}{\sin(x)}}{\frac{\sin^2(x) + \cos^2(x) - 2\sin^2(x)}{\sin(x)}} = \frac{\frac{1}{\sin(x)}}{\frac{\cos^2(x) - \sin^2(x)}{\sin(x)}}$$

Turning division into multiplication (multiplying the numerator by the reciprocal of the denominator)

$$\frac{1}{\sin(x)} \cdot \frac{\sin(x)}{\cos^2(x) - \sin^2(x)} = \frac{1}{\cos^2(x) - \sin^2(x)} = \frac{1}{\cos(2x)} = \sec(2x)$$

---

41. $\mathbf{csc(2x) = \frac{1}{2} sec(x) . csc(\mathit{x})}$

Solution:

Starting from the right-hand side:

$$\frac{1}{2}\sec(x).\csc(x) = \frac{1}{2}.\frac{1}{\cos(x)}.\frac{1}{\sin(x)} = \frac{1}{2\cos(x)\sin(x)}$$

$$= \frac{1}{\sin(2x)} = \csc(2x)$$

---

42. $\mathbf{sec(x)} = \frac{\sin(2x)}{\sin(x)} - \frac{\cos(2x)}{\cos(x)}$

Solution:

$$\frac{\sin(2x)}{\sin(x)} - \frac{\cos(2x)}{\cos(x)} =$$

The common denominator would be $\sin(x).\cos(x)$

$$\frac{\cos(x)\sin(2x)}{\sin(x).\cos(x)} - \frac{\sin(x)\cos(2x)}{\sin(x).\cos(x)}$$

$$= \frac{\cos(x)\,2\sin(x)\cos(x)}{\sin(x).\cos(x)} - \frac{\sin(x)\,(\cos^2(x) - \sin^2(x))}{\sin(x).\cos(x)}$$

Rearranging the first identity $1 - \cos^2(x) = \sin^2(x)$

$$= \frac{2\sin(x)\cos^2(x) - \sin(x)\,(\cos^2(x) - 1 + \cos^2(x))}{\sin(x).\cos(x)}$$

$$= \frac{2\sin(x)\cos^2(x) - \sin(x)\,(2\cos^2(x) - 1)}{\sin(x).\cos(x)}$$

Factor out sin(x)

$$= \frac{\sin(x)\,(2\cos^2(x) - 2\cos^2(x) + 1)}{\sin(x).\cos(x)} = \frac{1}{\cos(x)} = \sec(x)$$

4. Sine and Cosine Laws

4.1. Brief Introduction to Function

You can look at a function as a process in which you input something and you get something back (it could be the original thing). For example, you put seed in the ground, and you get a tree. The seed is your input, and the output would be the tree.

I do not intend to give you any more detailed information about function, but I just want to point out one other thing. For any given input, you will not get different outputs. That is, you plant a seed, and you will not get an apple tree and an orange tree at the same time. If you do, it is not a function. Let's put this into a little bit of a more abstract form. If "x" represent the inputs, and f(x) is a set of rule that defines the function (read f(x) as "f" of "x" or function of x). As an example, if we define a function in a way to double each input, it can be written as follows:

f(x)=2x that is if you input 10 you would get 20, and it should be noted as follows. F(10)=2(10)=20.

4.2. Sine Law

Let's talk about the area of a triangle. For that, we need a base and height. In a given triangle, we have three choices to pick these pairs. The base is usually easy to find; however, the height needs a bit of work. To locate the height, we may use the angle and length of the base.

Here is how. Let's say we have the following triangle:

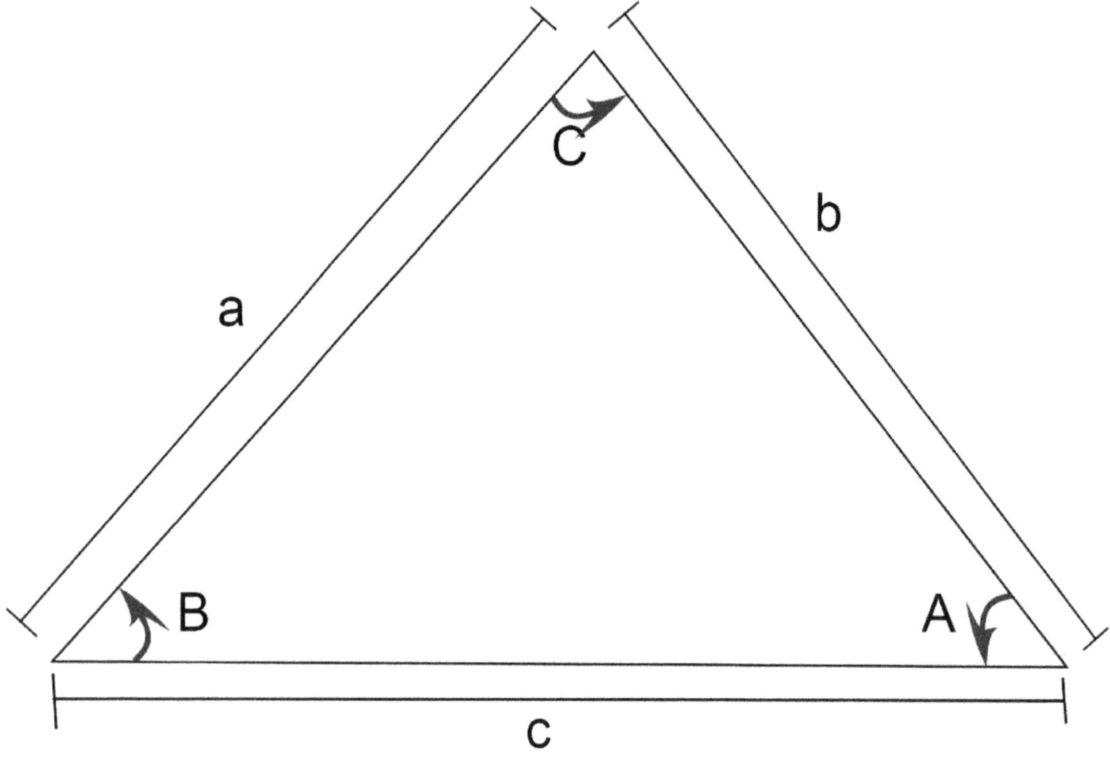

Figure 4.0.

We have three sides (triangle!) and three angles (triangle!). The point is that with any pair I choose to calculate the area, I must get the same result.

Let's take side c to be the base and try to find the associated height perpendicular to it. The next figure illustrates the aforementioned fact.

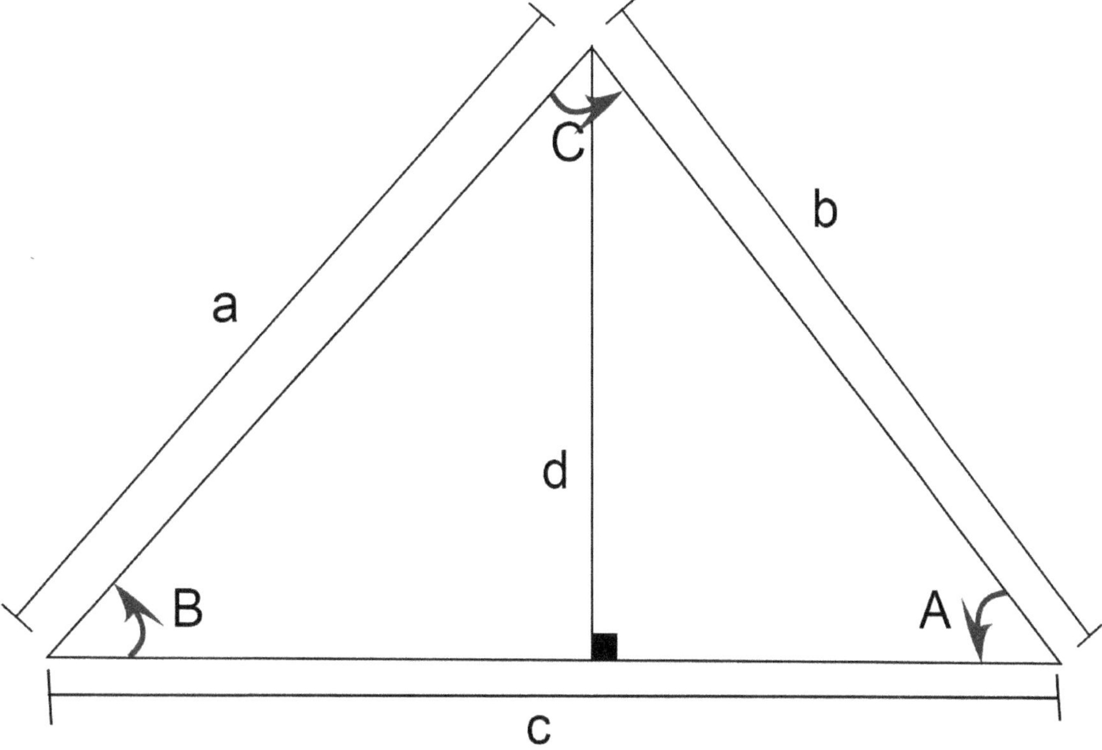

Figure 5.1.

Given that we have the length of sides and we know the angles, it is possible to find the height **d** of the triangle.

$$\sin(B) = \frac{d}{a} \Rightarrow d = a.\sin(B)$$

Therefore, the area of the triangle would be:

$$Area: \frac{1}{2}.c.a.\sin(B)$$

Now let's consider the same triangle, which was pictured before. This time, take the side "b" to be the base and find the height associated with it. As a result, the following graph is obtained.

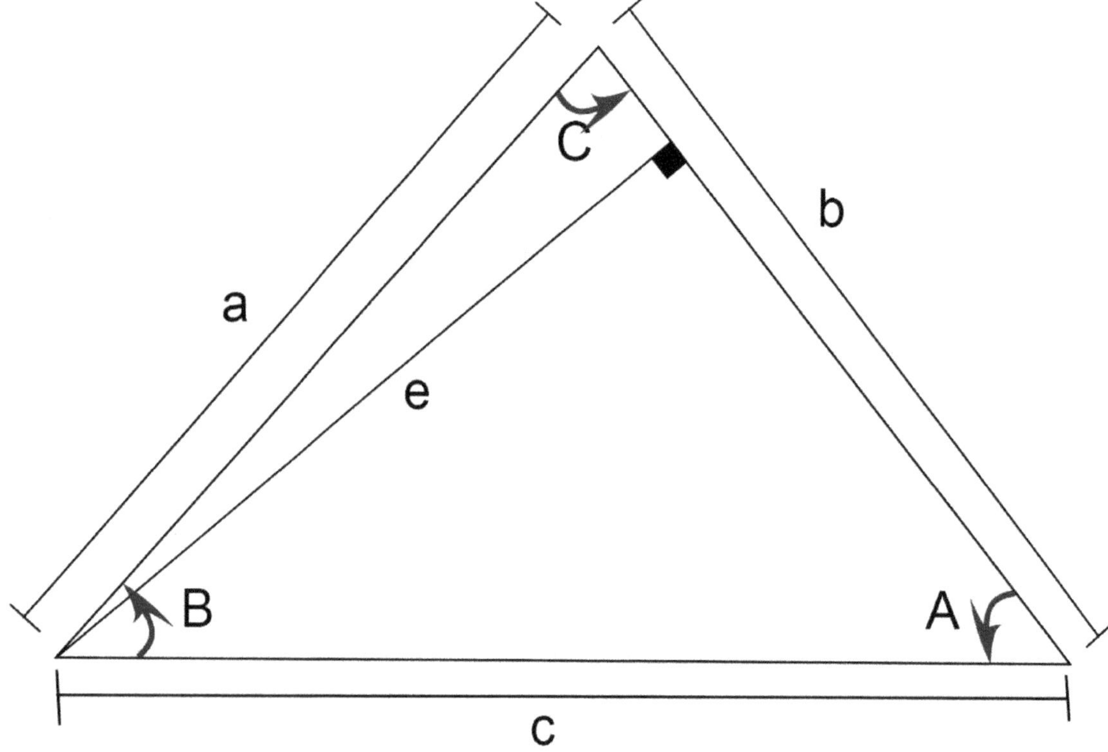

Figure 4.2.

Given that we have the length of the sides, and we know the angles, it is possible to find the height of the triangle.

$$\sin(A) = \frac{e}{c} => e = c.\sin(A)$$

Therefore, the area of the triangle would be:

$$Area: \frac{1}{2}.b.c.\sin(A)$$

Now let's consider the same triangle, which was pictured before. This time, take the side "a" to be the base and find the height associated with it. Therefore, the following graph is obtained.

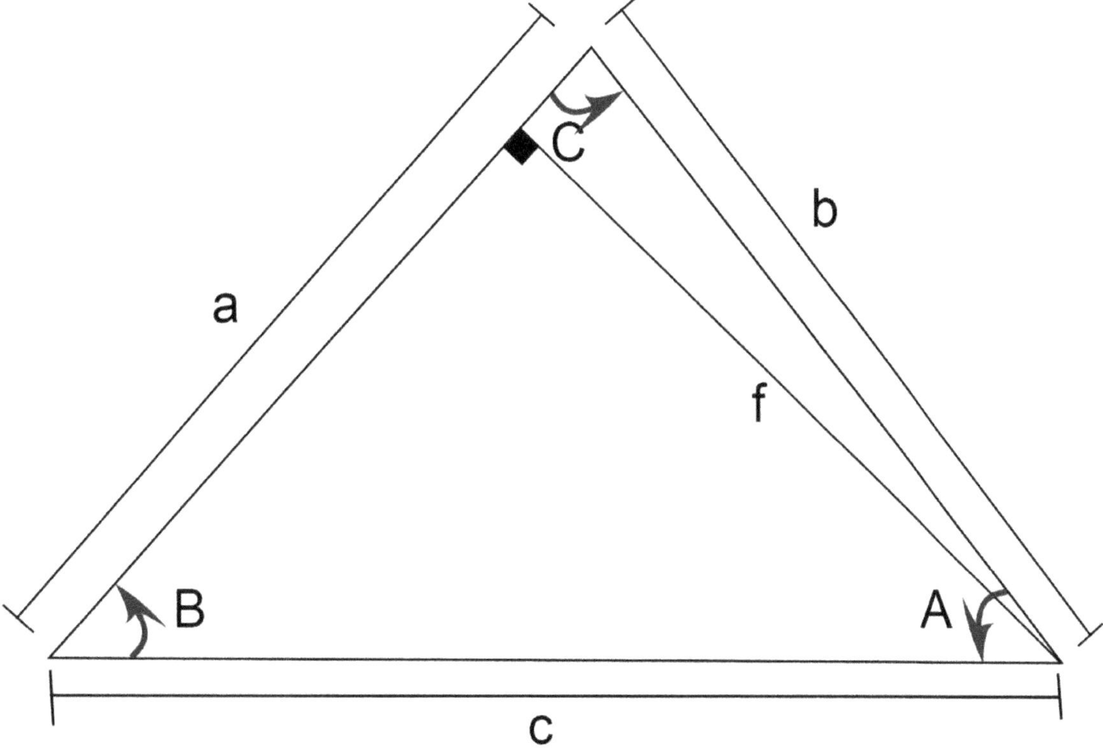

Figure 4.3.

Given that we have the length of the sides, and we know the angles, it is possible to find the height **f** of the triangle.

$$\sin(C) = \frac{f}{b} => f = b.\sin(C)$$

Ultimately, the area of the triangle would be:

$$Area: \frac{1}{2}.a.b.\sin(C)$$

The area of the triangle is equal regardless of the method we use. Hence, the following equalities hold:

Area: $\frac{1}{2}.a.b.\sin(C) = \frac{1}{2}.b.c.\sin(A) = \frac{1}{2}.c.a.\sin(B)$

$$\frac{1}{2}.a.b.\sin(C) = \frac{1}{2}.b.c.\sin(A) = \frac{1}{2}.c.a.\sin(B)$$

Multiply everything by 2.

$$a.b.\sin(C) = b.c.\sin(A) = c.a.\sin(B)$$

Now let's divide everything by a.b.c and see what we get:

$$\frac{a.b.\sin(C)}{a.b.c} = \frac{b.c.\sin(A)}{a.b.c} = \frac{c.a.\sin(B)}{a.b.c}$$

$$\frac{\sin(C)}{c} = \frac{\sin(A)}{a} = \frac{\sin(B)}{b}$$

The obtained result is rather cool! There are a lot of sines in it, so let's call it sine law!

This sine law is for the plane case. However, we do have sine law for hyperbolic, spherical and higher dimensions as well. Khawaja Nasir Al-din Al-Tusi, the Persian mathematician and philosopher, proved the sine law for the plane and spherical cases in the 13th century.

Examples

Given the following questions, determine the sine of missing angles and magnitude of the missing sides using sine law.

1. Given $\angle B = 107^{o}, q = 43, p = 28$ please find the value of $\sin(A)$ (final numerical value is not required).

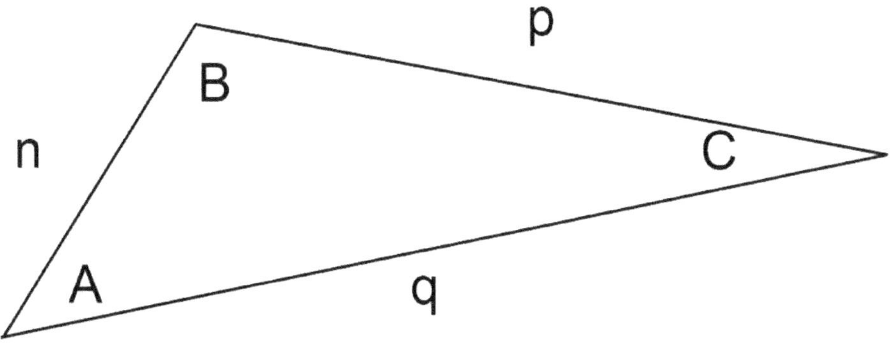

Figure 4.4.

Solution:

$$\frac{\boldsymbol{sin}(\boldsymbol{B})}{\boldsymbol{q}} = \frac{\boldsymbol{sin}(\boldsymbol{A})}{\boldsymbol{p}} = \frac{\boldsymbol{sin}(\boldsymbol{C})}{\boldsymbol{n}}$$

$$=> \frac{\boldsymbol{sin}(107^o)}{\mathbf{43}} = \frac{\boldsymbol{sin}(\boldsymbol{A})}{\mathbf{28}} = \frac{\boldsymbol{sin}(\boldsymbol{C})}{\boldsymbol{n}}$$

$$=> \frac{\boldsymbol{sin}(107^o)}{\mathbf{43}} = \frac{\boldsymbol{sin}(\boldsymbol{A})}{\mathbf{28}} => \frac{\mathbf{28}.\boldsymbol{sin}(107^o)}{\mathbf{43}} = \boldsymbol{sin}(\boldsymbol{A})$$

---

2. Given $\sin(A) = 0.5, q = 12, p = 12$ evaluate the length of n with respect to sin (B).

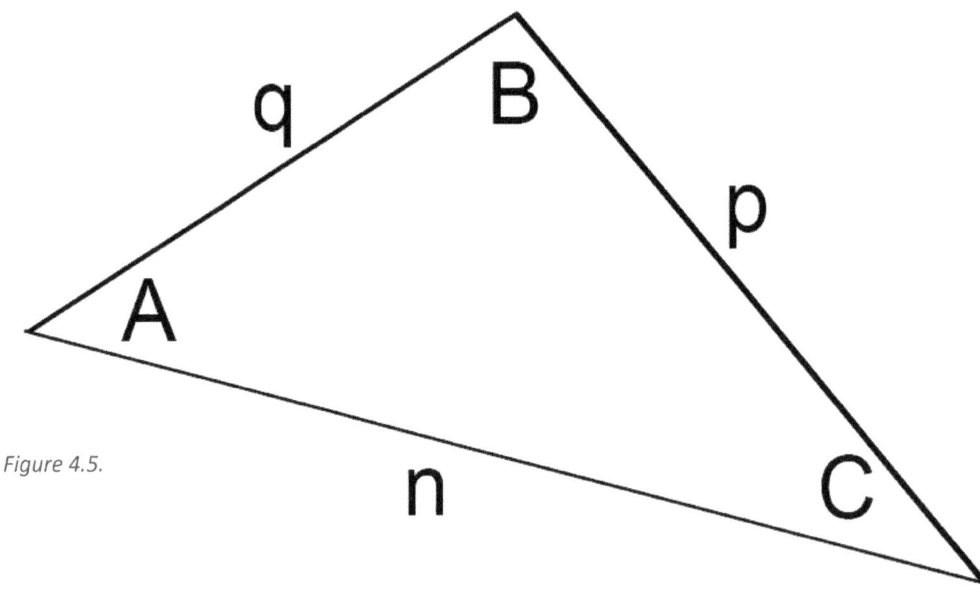

Figure 4.5.

Solution:

$$\frac{sin(B)}{n} = \frac{sin(A)}{p} = \frac{sin(C)}{q} => \frac{sin(B)}{n} = \frac{0.5}{12} = \frac{sin(C)}{12}$$

Since sides p and q have the same length, it suffices to say they have equal angles. That is, sin(C)=Sin(A)=0.5

$$=> \frac{sin(B)}{n} = \frac{0.5}{12} = \frac{0.5}{12}$$

To find the value of sin (B), we can use the following logic:

Since sin(A)=.5, then we can conclude ∠A =30°=∠C.

The sum of interior angles of a triangle adds up 180°; therefore, the following can be found:

$$\angle B = 180^o - 30^o - 30^o = 120^o$$

$$\sin(120^o) = \frac{\sqrt{3}}{2}$$

$$=> \frac{\frac{\sqrt{3}}{2}}{n} = \frac{0.5}{12} = \frac{0.5}{12} => \frac{\frac{\sqrt{3}}{2}}{n} = \frac{0.5}{12}$$

Flip both sides of the equality (since we know n cannot be zero).

$$=> \frac{n}{\frac{\sqrt{3}}{2}} = \frac{12}{0.5} => n = \frac{12\sqrt{3}}{2(0.5)} = 12\sqrt{3}$$

Length of n is $12\sqrt{3}$

4.3. Cosine Law (Kashi's Law)

Let's focus on the following triangle. We are trying to find the length of the side j using the height d and the angle B and A.

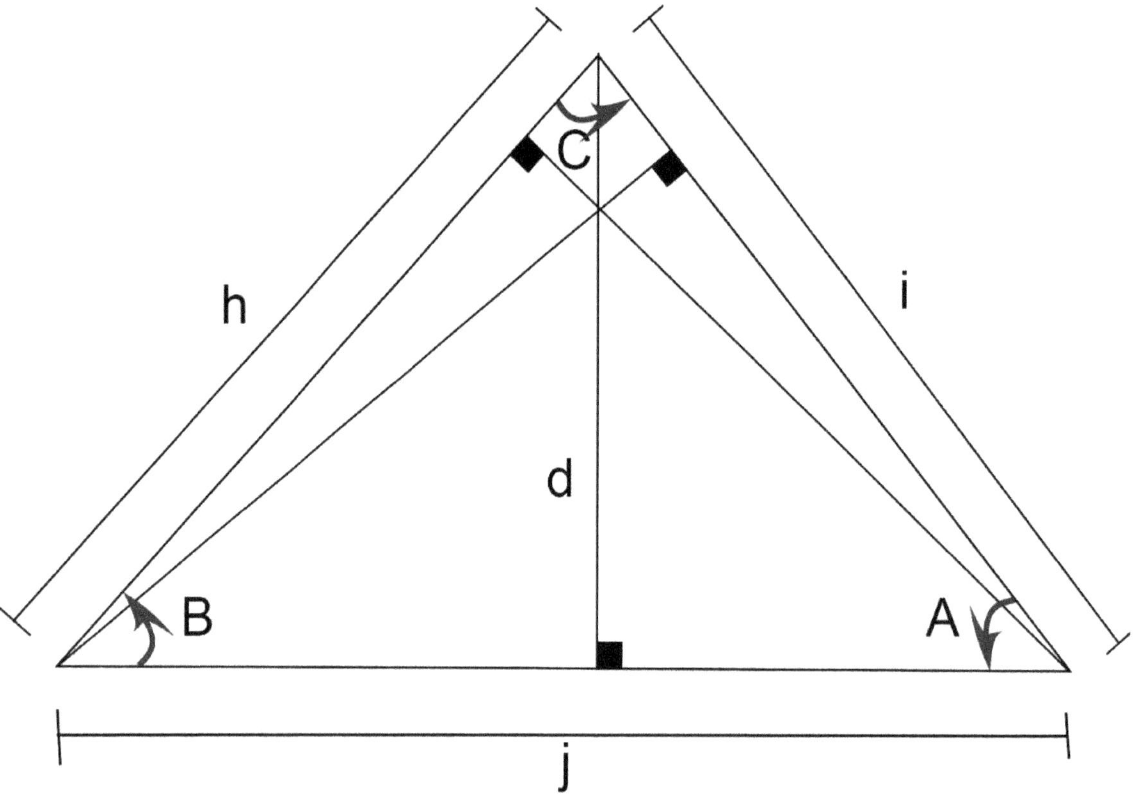

Figure 4.6.

Notice the length of j consists of two sides. We can add the associated lengths to find the length of j.

That is, in the following graph, the length of c + e is equal to j.

Considering the length c, would the following make sense?

$$\cos(B) = \frac{c}{h} => h.\cos(B) = c$$

$$\cos(A) = \frac{e}{i} => i.\cos(A) = e$$

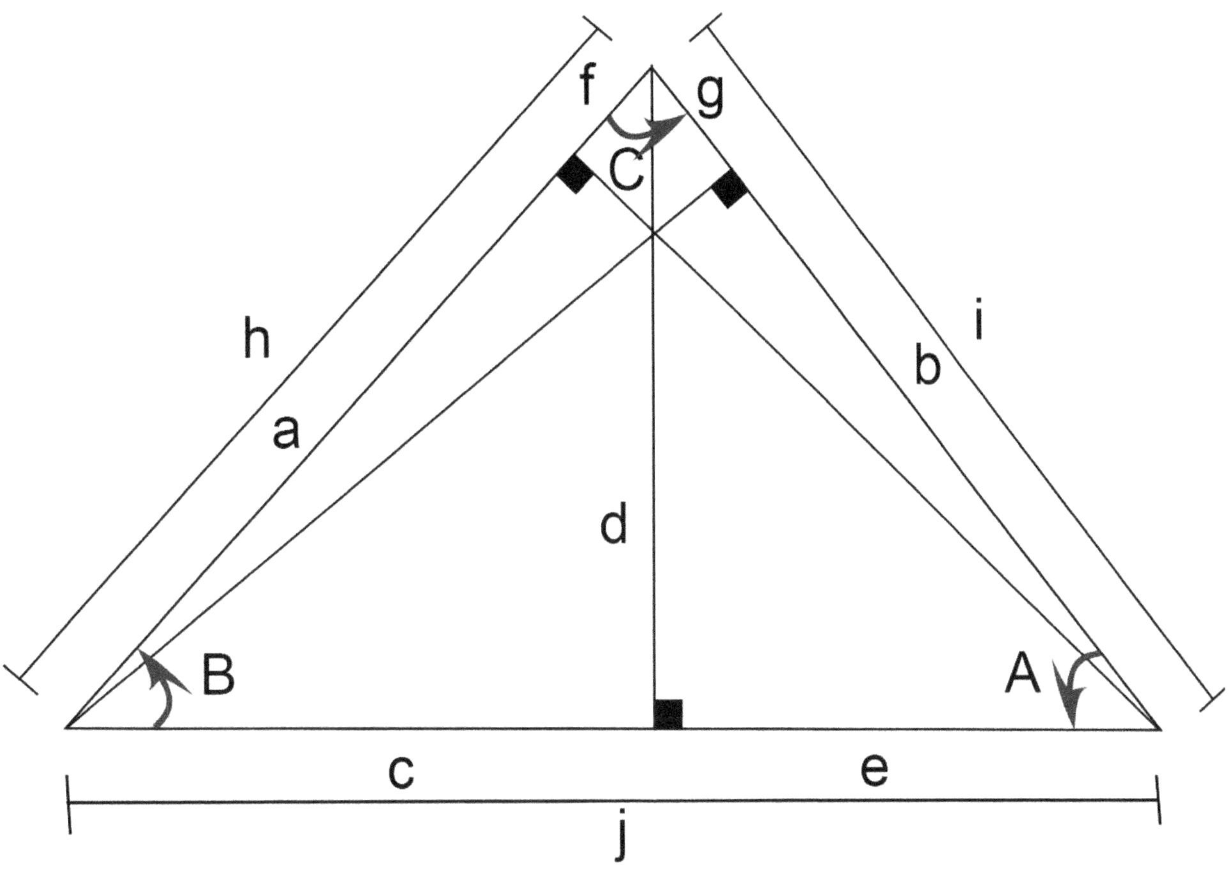

Figure 4.7.

Therefore, the following can be obtained.

$$j = c + e => j = h.\cos(B) + i.\cos(A)$$

Multiply everything by j, and the following can be obtained:

$$j^2 = j.h.\cos(B) + j.i.\cos(A)$$

Looking at side h, we can conclude the following:

$$h = a + f$$

$$\cos(B) = \frac{a}{j} => j.\cos(B) = a$$

$$\cos(C) = \frac{f}{i} => i.\cos(C) = f$$

Accordingly, the following can be obtained:

$$h = a + f => h = j.\cos(B) + i.\cos(C)$$

Multiply everything by h, and the following can be achieved:

$$h^2 = h.j.\cos(B) + h.i.\cos(C)$$

Looking at side "I," we can conclude the following:

$$i = g + b$$

$$\cos(A) = \frac{b}{j} => j.\cos(A) = b$$

$$\cos(C) = \frac{g}{h} => h.\cos(C) = g$$

Therefore, the following can be achieved:

$$i = g + b => i = h.\cos(C) + j.\cos(A)$$

Multiply everything by "i", so we can achieve the following:

$$i^2 = i.h.\cos(C) + i.j.\cos(A)$$

Let's now add up the obtained results for i^2, and j^2 then subtract h^2 from it to get the following:

$j^2 = j.h.\cos(B) + j.i.\cos(A)$

$h^2 = h.j.\cos(B) + h.i.\cos(C)$

$i^2 = i.h.\cos(C) + i.j.\cos(A)$

$=> j^2 + i^2 - h^2 = j.h.\cos(B) + j.i.\cos(A) +$

$i.h.\cos(C) + i.j.\cos(A) - h.j.\cos(B) - h.i.\cos(C)$

$=> j^2 + i^2 - h^2 = j.i.\cos(A) + i.j.\cos(A)$

$=> j^2 + i^2 - h^2 = 2j.i.\cos(A)$

Rearranging

$=> j^2 + i^2 - 2j.i.\cos(A) = h^2$

O! Is it cosine law? I guess it is :D

There are many ways you can prove cosine law, and we just used one.

Thus, cosine law for any arbitrary triangle can be formulated by the three equations. That is, given we know the length of two sides of any given triangle and the angle between the two sides, we can find the length of the other side using Kashi's law. Please note if we have the magnitude of other angles, it is possible to obtain the angles between the sides using Kashi's law.

As for the next triangle, cosine law can be written as follows.

$$j^2 + i^2 - 2j.i.\cos(A) = h^2$$

And

$$h^2 + i^2 - 2h.i.\cos(C) = j^2$$

And

$$j^2 + h^2 - 2j.h.\cos(B) = i^2$$

In simple terms: in a given triangle, the length of each side squared is equal to the addition of the square of the other sides' lengths minus the multiplication of the other sides' lengths with the cosine of the angle between the other sides. It was not simple terms at all. I like the formula more :D.

Please note that the cosine law, unlike Pythagorean law, does not require any right-angled triangle. However, we can conclude the Pythagorean law is a special case for cosine law.

That is, imagine side j is perpendicular to side i. Then the following should make sense.

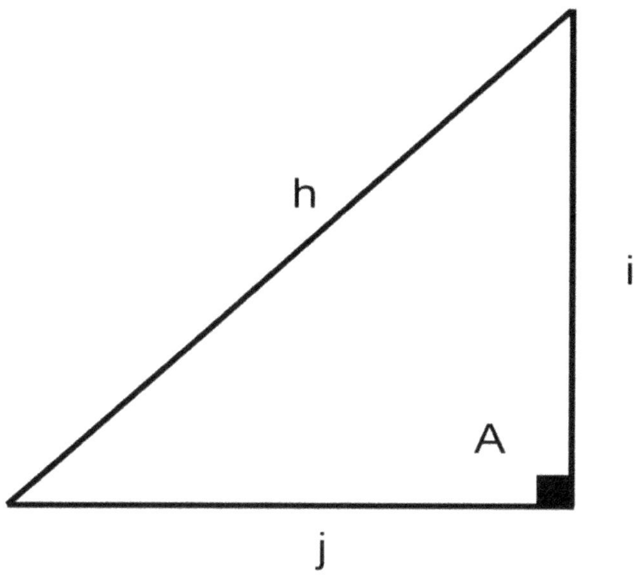

Figure 4.8.

$h^2 = j^2 + i^2 - 2j.i.\cos(A) = j^2 + i^2 - 2j.i.\cos(90^o)$

$= j^2 + i^2 - 2j.i.0 = j^2 + i^2$

$=> h^2 = j^2 + i^2$

Which is Pythagorean theorem. Therefore, it suffices to say that Pythagorean theorem is a special case for Cosine (Kashi's) Law.

Example

Find the length of side "c" using the cosine law given the length of sides "a" and "b" are 12, and 11, respectively, and the angle A is 30°.

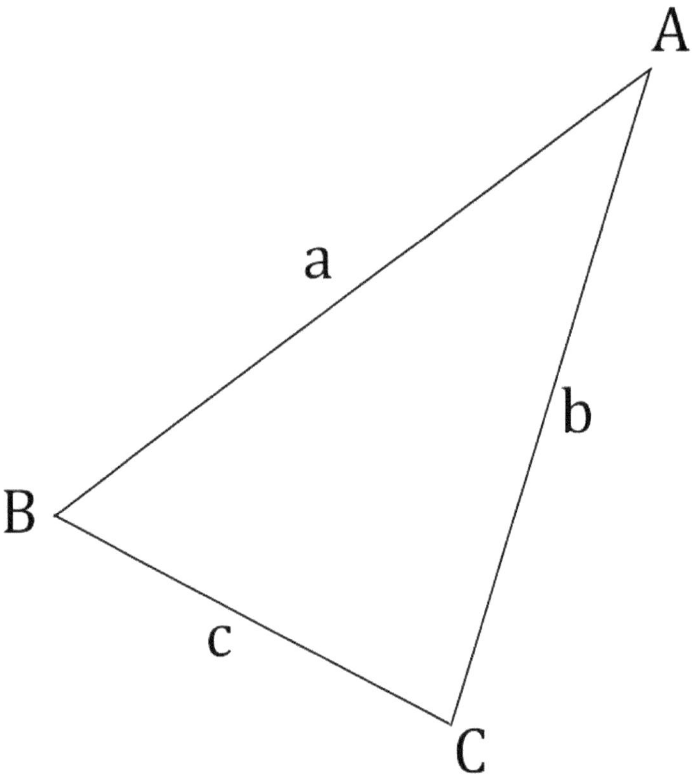

Figure 4.9.

Solution:

$a^2 + b^2 - b.a.\cos(A) = c^2$

$$12^2 + 11^2 - 11.12.\cos(30) = c^2 \Rightarrow 144 + 121 + \frac{132\sqrt{3}}{2} = 265 + 66\sqrt{3}$$

5. Graphing Trigonometric Functions

5.1. Simple Way to Graph Sin(x)

To graph sin(x) or cos(x), one easy way would be to draw two perpendicular lines (a coordinate system). The vertical line can be your function (for example y = sin(x)), and the horizontal axis can show the magnitude of the associated angles, which is the x value. The unit for the horizontal axis, which are the x values, can be in radian. However, you may try any other units of angles, such as degrees.

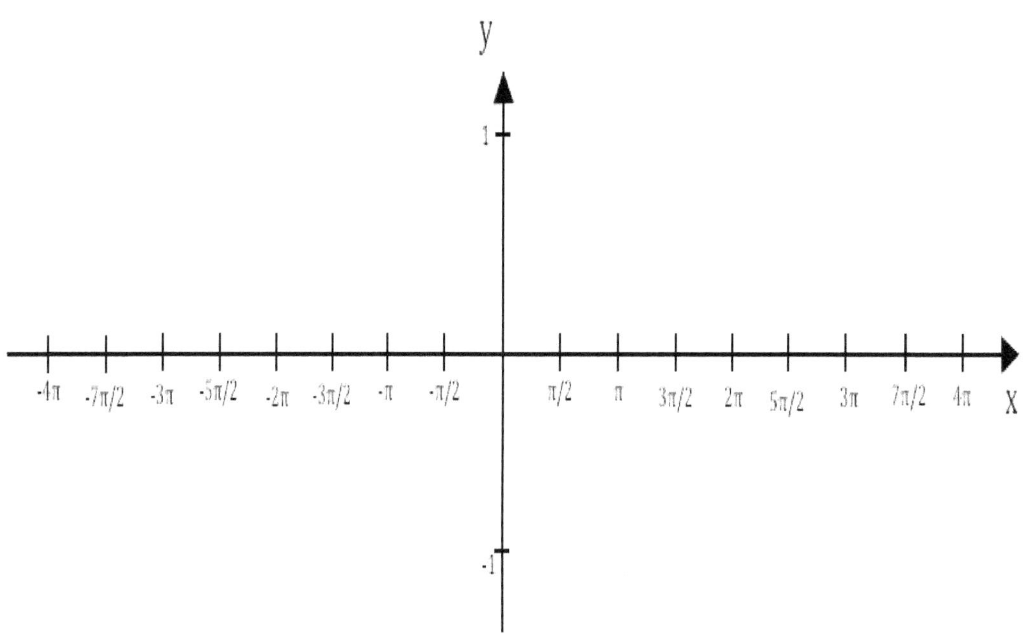

Figure 6.1.

Initially, we are going to graph y=sin(x) in a very simple fashion by finding the coordinates of the point associated to sin(x). To do so, try to find values of sin(x) for a few critical angles. You may find the values of the associated points from the table in chapter one, which we constructed.

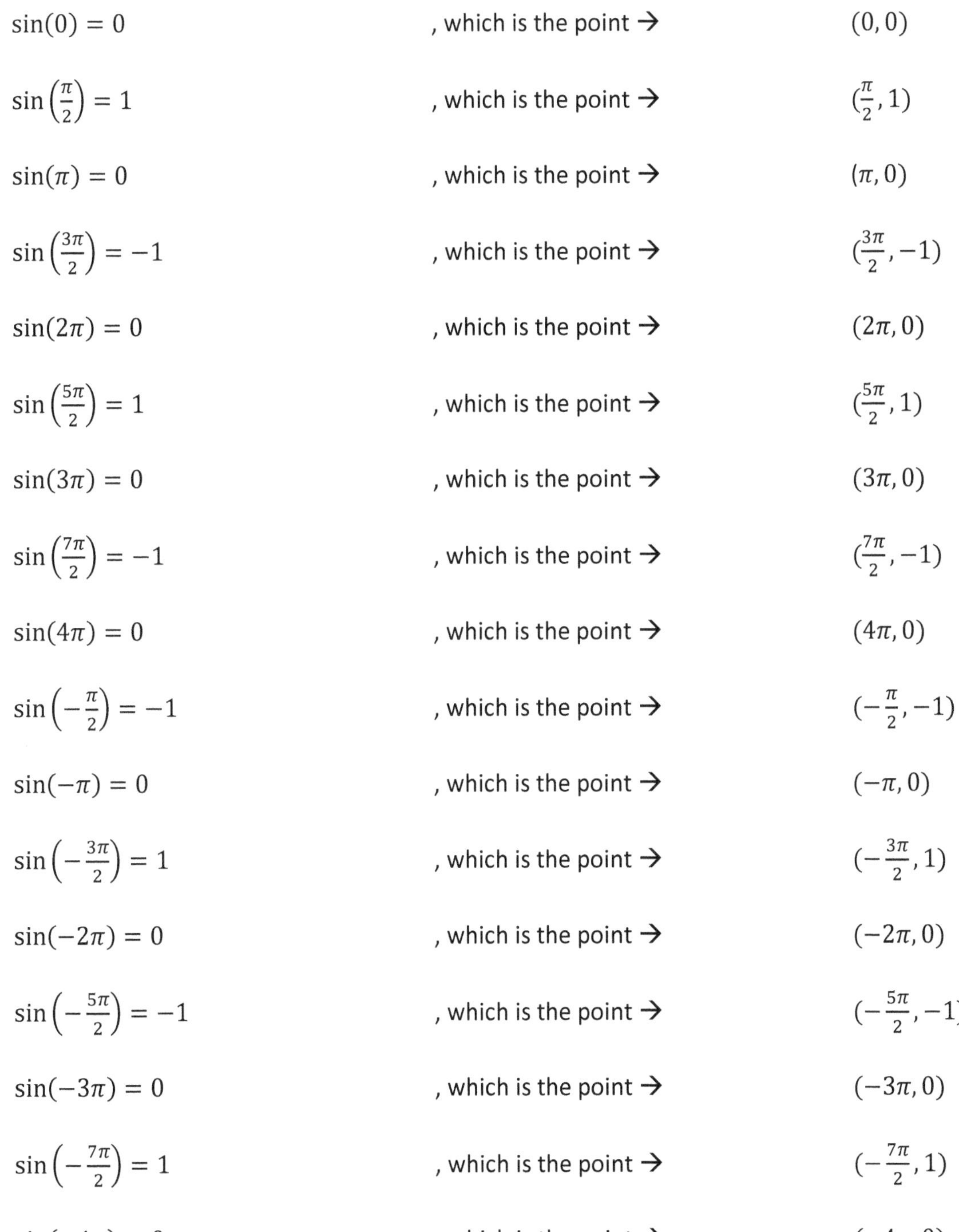

$\sin(0) = 0$	, which is the point →	$(0, 0)$
$\sin\left(\frac{\pi}{2}\right) = 1$	, which is the point →	$(\frac{\pi}{2}, 1)$
$\sin(\pi) = 0$	, which is the point →	$(\pi, 0)$
$\sin\left(\frac{3\pi}{2}\right) = -1$	, which is the point →	$(\frac{3\pi}{2}, -1)$
$\sin(2\pi) = 0$	, which is the point →	$(2\pi, 0)$
$\sin\left(\frac{5\pi}{2}\right) = 1$	, which is the point →	$(\frac{5\pi}{2}, 1)$
$\sin(3\pi) = 0$	, which is the point →	$(3\pi, 0)$
$\sin\left(\frac{7\pi}{2}\right) = -1$	, which is the point →	$(\frac{7\pi}{2}, -1)$
$\sin(4\pi) = 0$	, which is the point →	$(4\pi, 0)$
$\sin\left(-\frac{\pi}{2}\right) = -1$	, which is the point →	$(-\frac{\pi}{2}, -1)$
$\sin(-\pi) = 0$	, which is the point →	$(-\pi, 0)$
$\sin\left(-\frac{3\pi}{2}\right) = 1$	, which is the point →	$(-\frac{3\pi}{2}, 1)$
$\sin(-2\pi) = 0$	, which is the point →	$(-2\pi, 0)$
$\sin\left(-\frac{5\pi}{2}\right) = -1$	, which is the point →	$(-\frac{5\pi}{2}, -1)$
$\sin(-3\pi) = 0$	, which is the point →	$(-3\pi, 0)$
$\sin\left(-\frac{7\pi}{2}\right) = 1$	, which is the point →	$(-\frac{7\pi}{2}, 1)$
$\sin(-4\pi) = 0$	, which is the point →	$(-4\pi, 0)$

Now try to find the aforementioned points on the coordinate system that you have created.

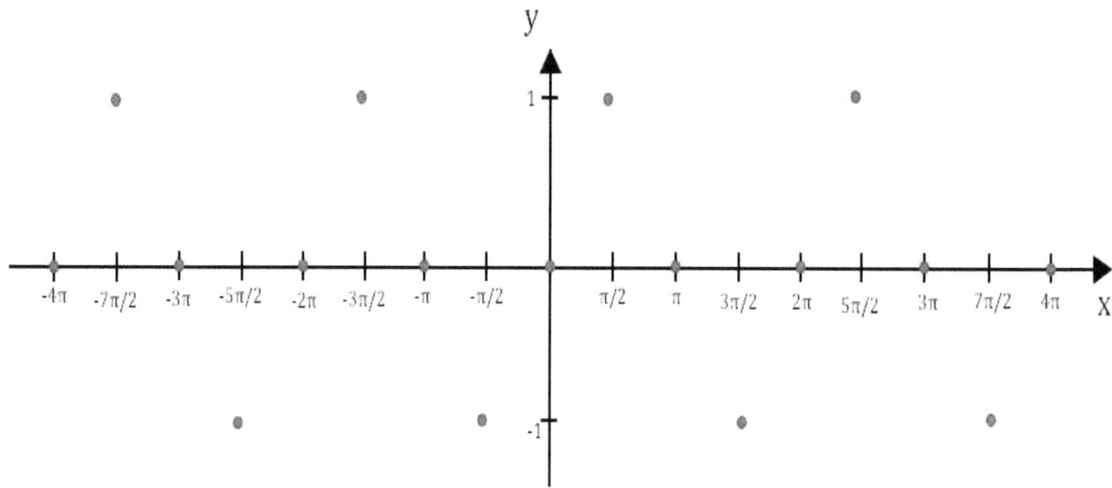

Figure 5.2.

Now, by connecting the dots in a smooth way, you would get the following shape for sin(x). I am completely aware I have not talked about what I mean by smooth or the proper way of connecting dots. Just try to be gentle :D.

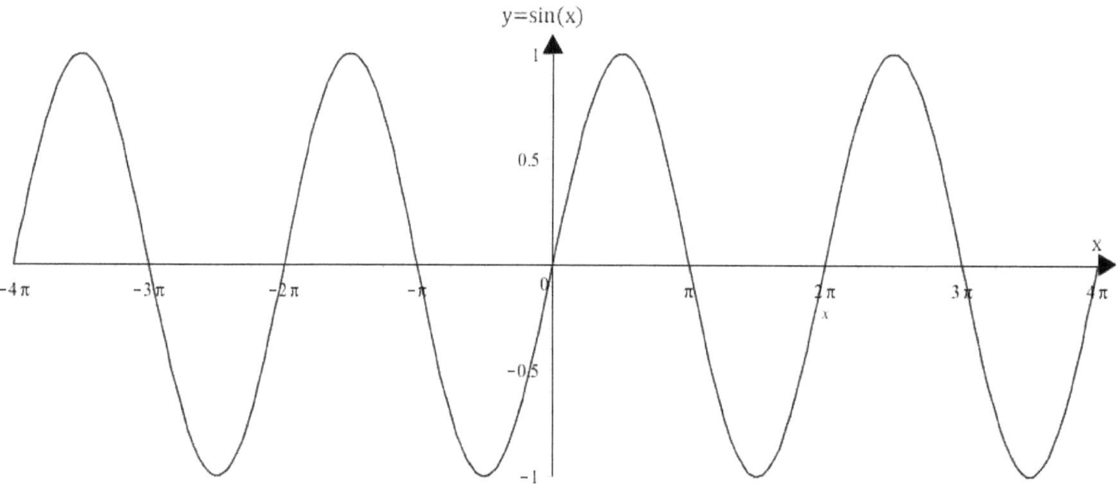

Figure 5.3.

5.2. Simple Way to Graph Cos(X)

As for the sin(x), first we are going to graph y=cos(x) in a very simple fashion by finding the coordinates of the point associated with cos(x). To do so, try to find values of cos(x) for a few critical angles. You may find the values of the associated points from the table in chapter one, which we constructed.

$\cos(0) = 1$, which is the point → $(0, 1)$

$\cos\left(\frac{\pi}{2}\right) = 0$, which is the point → $(\frac{\pi}{2}, 0)$

$\cos(\pi) = -1$, which is the point → $(\pi, -1)$

$\cos\left(\frac{3\pi}{2}\right) = 0$, which is the point → $(\frac{3\pi}{2}, 0)$

$\cos(2\pi) = 1$, which is the point → $(2\pi, 1)$

$\cos\left(\frac{5\pi}{2}\right) = 0$, which is the point → $(\frac{5\pi}{2}, 0)$

$\cos(3\pi) = -1$, which is the point → $(3\pi, -1)$

$\cos\left(\frac{7\pi}{2}\right) = 0$, which is the point → $(\frac{7\pi}{2}, 0)$

$\cos(4\pi) = 1$, which is the point → $(4\pi, 1)$

$\cos\left(-\frac{\pi}{2}\right) = 0$, which is the point → $(-\frac{\pi}{2}, 0)$

$\cos(-\pi) = -1$, which is the point → $(-\pi, -1)$

$\cos\left(-\frac{3\pi}{2}\right) = 0$, which is the point → $(-\frac{3\pi}{2}, 0)$

$\cos(-2\pi) = 1$, which is the point → $(-2\pi, 1)$

$\cos\left(-\frac{5\pi}{2}\right) = 0$, which is the point → $(-\frac{5\pi}{2}, 0)$

$\cos(-3\pi) = -1$, which is the point → $(-3\pi, -1)$

$\cos\left(-\frac{7\pi}{2}\right) = 0$, which is the point → $(-\frac{7\pi}{2}, 0)$

$\cos(-4\pi) = 1$, which is the point → $(-4\pi, 1)$

Next step is to find the mentioned points on the coordinate system as follow:

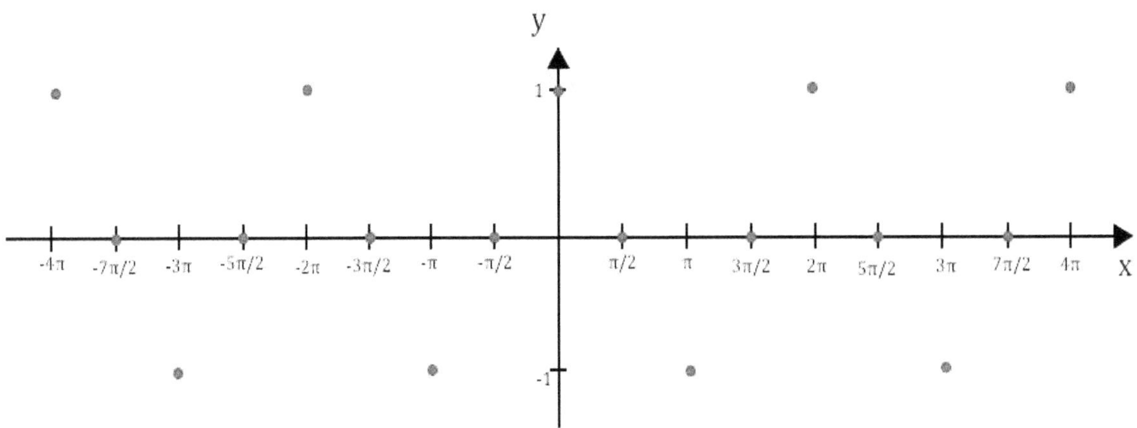

Figure 5.4.

Ultimately, you would get the following graph by connecting the dots in a smooth manner.

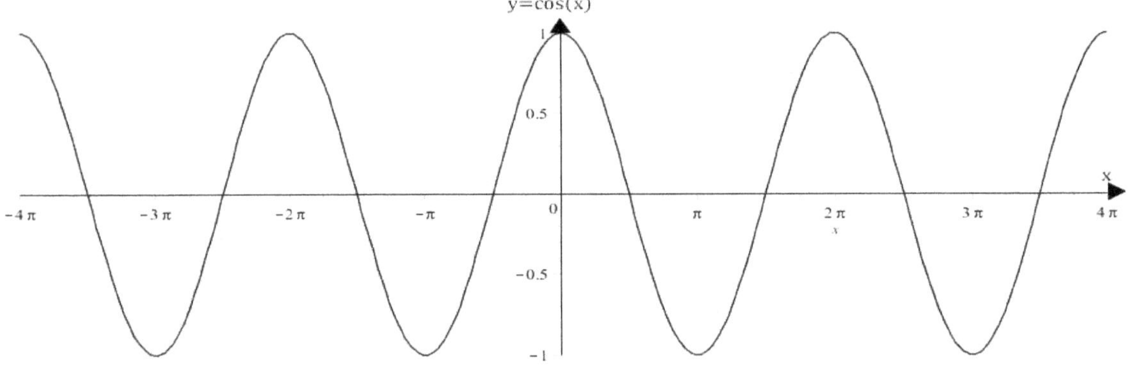

Figure 5.5.

5.3. Amplitude, Period, Peak, Crest and Median Line

Hopefully, the graph of y = sin(x) and y= cos(x) makes sense. Next task is to try to find how we can transform them. That is moving them and stretching them. To do so, we need to define a few terms, which can come handy in many cases.

Median line: a horizontal line that passes through the middle of the sin(x) or cos(x) function. In the following graph, it would be the x-axis (or y=0).

Peak or crest: The highest point of the sin(x) or cos(x) function.

Trough: The lowest point of the sin(x) or cos(x) function.

Amplitude: The distance from median line to the peak/crest (highest point) or distance from median line to trough.

Period, wavelength: One complete cycle. That is, you start from a certain point (y-value) and a particular direction, and you get back to the same point while keeping the same direction.

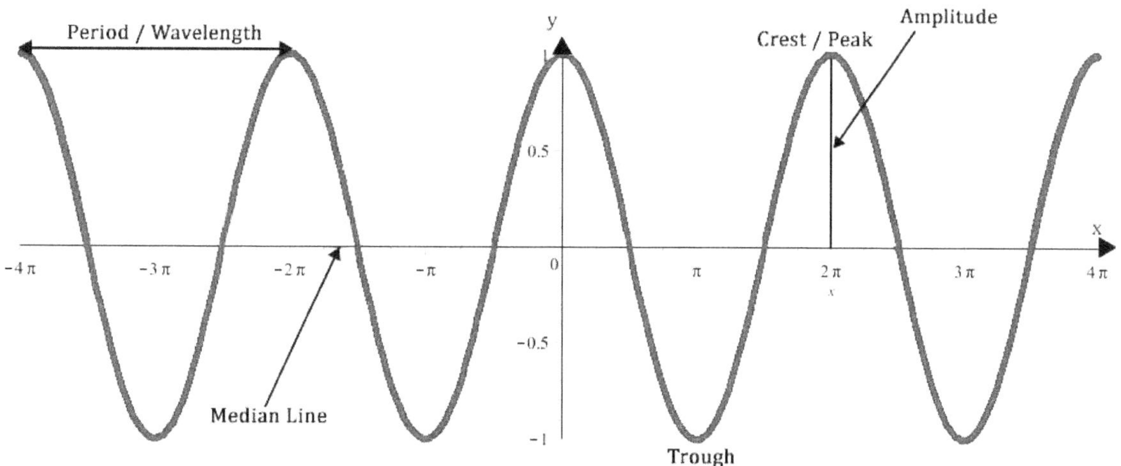

Figure 5.6.

Determine the amplitude in the following sine example:

1. $y = \sin(x)$

Solution:

We have to look at "A," which is multiplied by sine function as it is in the following format:

$y = A\sin(k(x - d)) + c$

That is, amplitude = 1

---

2. $y = 2\sin(x)$

Solution:

Amplitude = 2

---

3. $y = \frac{1}{5}\sin(x)$

Solution:

Amplitude = $\frac{1}{5}$

---

4. $y = -3\sin(x)$

Solution:

Amplitude =-3

---

5. $y = .67 \sin(x)$

Solution:

Amplitude = 0.67

---

6. $y = 4\sin(3x)$

Solution:

Amplitude = 4

---

7. $y = 3\sin(2x)$

Solution:

Amplitude = 3

---

8. $y = 213 \sin\left(\frac{x}{6}\right)$

Solution:

Amplitude = 213

---

9. $y = 14\sin(x + 4\pi)$

Solution:

Amplitude =14

---

10. $3y = 2\sin(x - 4)$

Solution:

$3y = 2\sin(x - 4) \Rightarrow \textit{devide both side by } 3 \Rightarrow \frac{3y}{3} = \frac{2\sin(x - 4)}{3}$

$\Rightarrow y = \frac{2\sin(x - 4)}{3} \Rightarrow y = \frac{2}{3}\sin(x - 4)$

Amplitude $= \frac{2}{3}$

---

11. $y = \frac{-2\sin(2x-3)}{3}$

Solution:

$y = \frac{-2\sin(2x - 3)}{3} \Rightarrow y = \frac{-2}{3}\sin(2x - 3)$

Amplitude $= -\frac{2}{3}$

---

12. $\frac{y}{5} = 2\sin(x)$

Solution:

$\frac{y}{5} = 2\sin(x) => \textit{multiply both sides by } 5 => 5 \times \frac{y}{5} = 5 \times 2\sin(x)$

$=> y = 10\sin(x)$

Amplitude = 10

---

13. $\frac{y}{3} = \frac{2}{3}\sin(5x)$

Solution:

$\frac{y}{3} = \frac{2}{3}\sin(5x) => \textit{multiply both sides by } 3 => 3 \times \frac{y}{3} = 3 \times \frac{2}{3}\sin(5x) =>$

$y = 2\sin(5x)$

Amplitude = 2

---

14. $\frac{y}{7} + 3 = 2\sin\left(\frac{x}{5}\right) - 6$

Solution:

$\frac{y}{7} + 3 = 2\sin\left(\frac{x}{5}\right) - 6 => \textit{multiply both sides by } 7$

$=> 7 \times (\frac{y}{7} + 3) = 7 \times (2\sin\left(\frac{x}{5}\right) - 6)$

$=> y + 21 = 14\sin\left(\frac{x}{5}\right) - 42$ Moving 21 to the other side would make the standard form, but it does not have any effect on the result of amplitude:

$y + 21 - 21 = 14\sin\left(\frac{x}{5}\right) - 42 - 21 =>$

$y = 14\sin\left(\frac{x}{5}\right) - 63$

Amplitude = 14

---

15. $y = 4\sin(x^2 + 2)$

Solution:

Amplitude =4

---

16. $\frac{y}{4} = 2\sin(x^2 - x)$

Solution:

$\frac{y}{4} = 2\sin(x^2 - x) => multiply\ both\ sides\ by\ 4$

$=> 4 \times \frac{y}{4} = 4 \times 2\sin(x^2 - x)$

$=> y = 8\sin(x^2 - x)$

Amplitude = 8

---

17. $y = 3\sin(x) + 2$

Solution:

Amplitude = 3

---

18. $y = \sin\left(3x - \frac{3\pi}{4}\right)$

Solution:

Amplitude = 1

---

19. $2y = 5\sin\left(\frac{x}{3} + 12\right) - 4$

Solution:

$2y = 5\sin\left(\frac{x}{3} + 12\right) - 4 \Rightarrow \textit{divide both sides by } 2 \Rightarrow \frac{2y}{2} = \frac{(5\sin\left(\frac{x}{3}+12\right)-4)}{2}$

$\Rightarrow y = \frac{5}{2}\sin\left(\frac{x}{3} + 12\right) - \frac{4}{2}$

Amplitude = $\frac{5}{2}$

---

20. $2y = \frac{12}{5}\sin(x - 2)$

Solution:

$$2y = \frac{12}{5}\sin(x-2) \Rightarrow \textit{divide both sides by } 2 \Rightarrow \frac{2y}{2} = \frac{\frac{12}{5}\sin(x-2)}{2}$$

$$\Rightarrow y = \frac{12}{10}\sin(x-2) = \frac{6}{5}\sin(x-2)$$

Amplitude = $\frac{6}{5}$

5.4. Period of sin(x) and cos(x)

As you might have noticed, the period of the sin(x) and cos(x) can be obtained from the graphs shown, which is equal to 2π. That is, for every 2π, the sin(x) and cos(x) function would repeat themselves. Please refer to the following illustrations for further clarification:

y = cos(x)

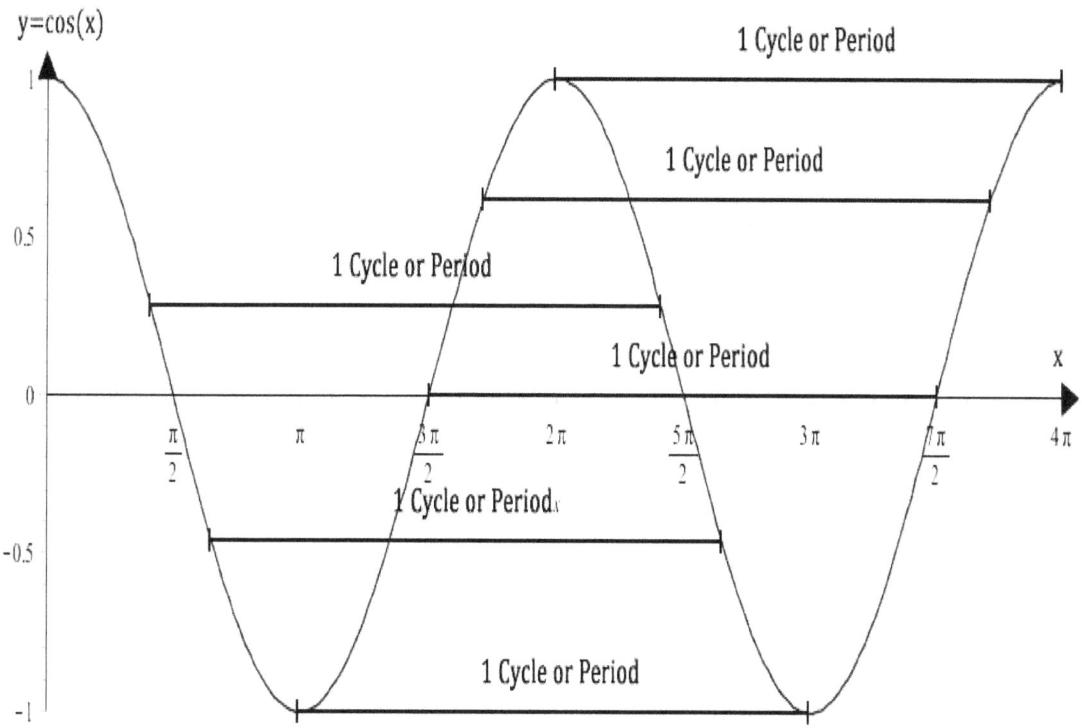

Figure 5.7.

y = sin(x)

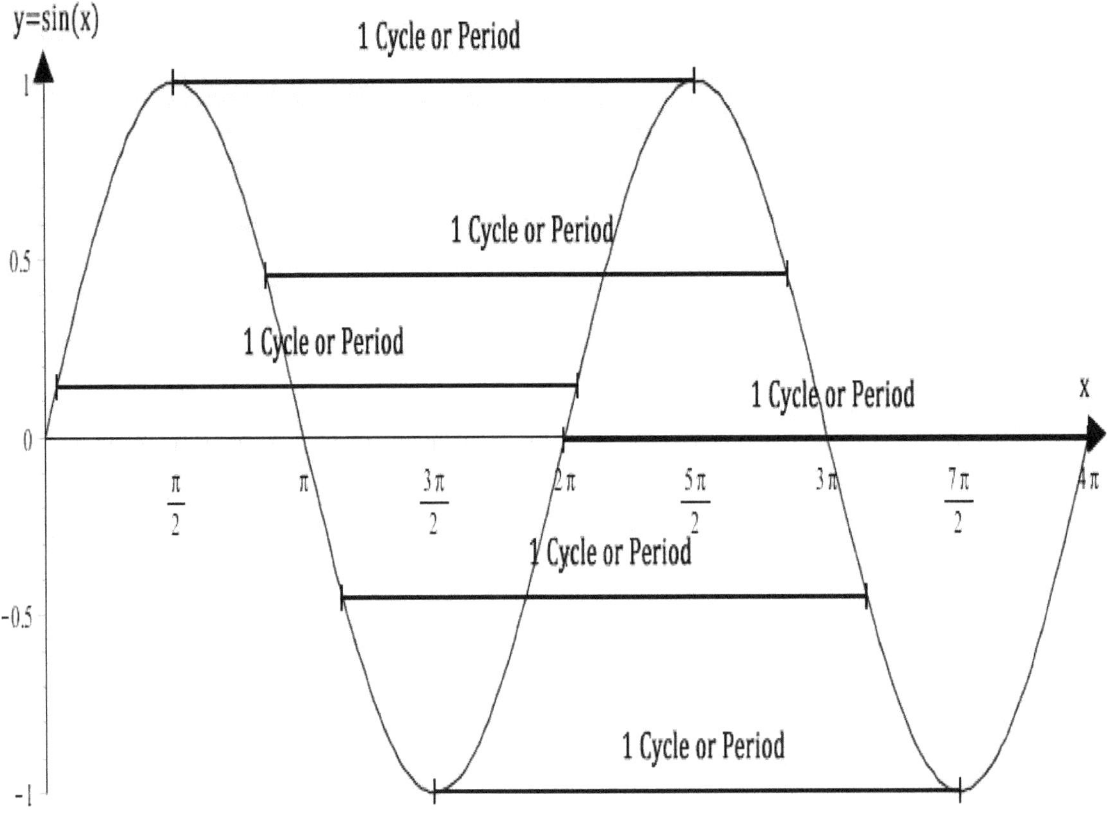

Figure 5.8.

Examples

Find the period of the following graphs:

1. $y = sin\ (2x)$

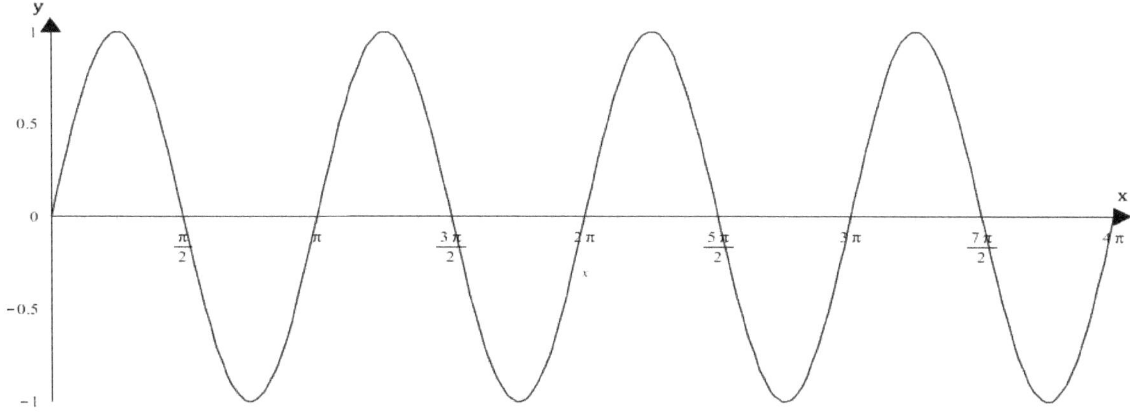

Figure 5.9.

Solution:

The period is π since there is one cycle from 0 to π (please refer to the graph for better visualization).

---

2. $y = sin\,(\frac{x}{2})$

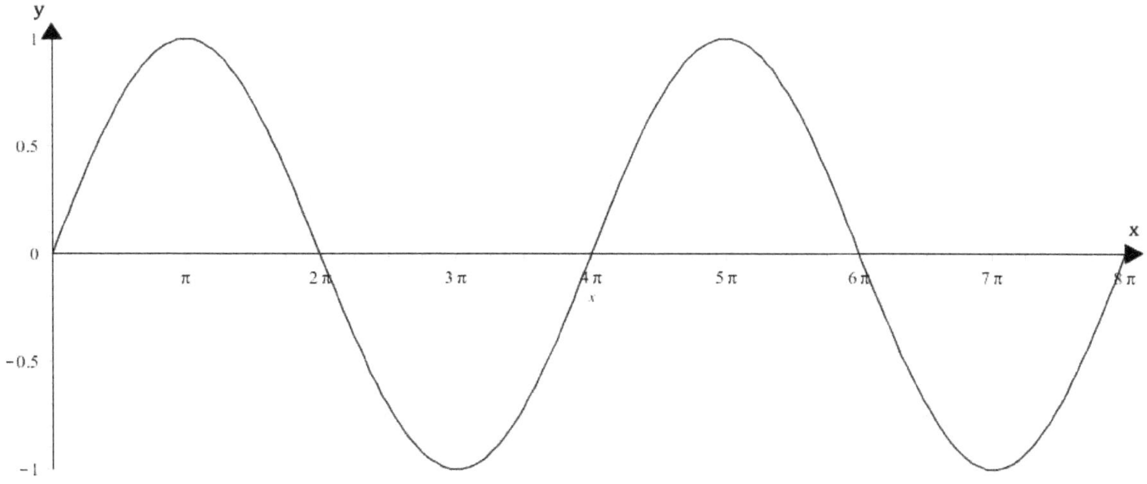

Figure 5.10.

4π

---

3. $y = sin\ (\pi x)$

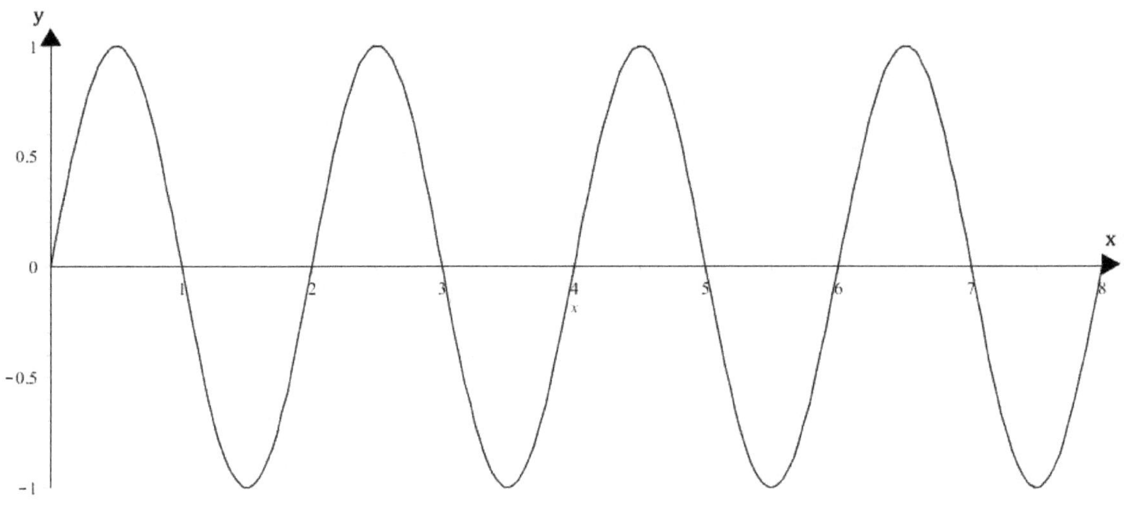

Figure 5.11.

Solution:

2

---

4. $y\ = \cos(6(x + \pi))$

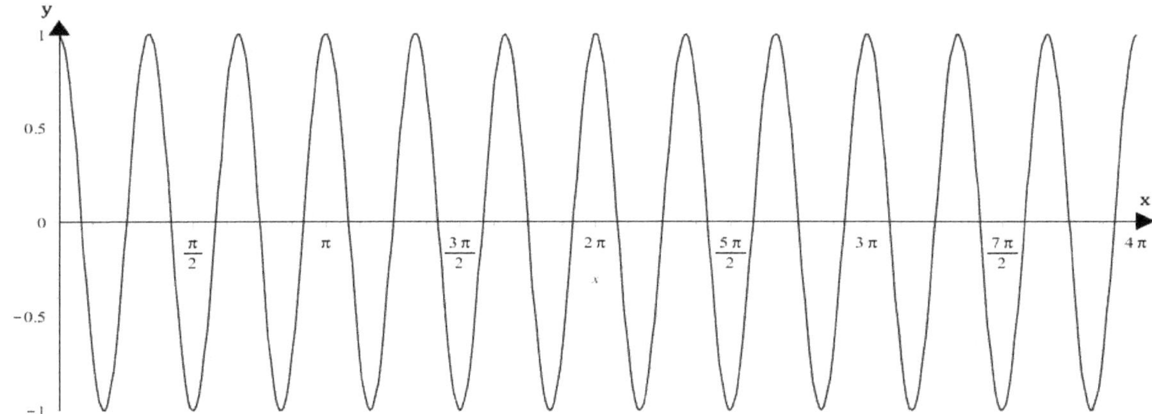

Figure 5.12.

Solution:

$\frac{\pi}{3}$, How? As you can see from 0 to π, we have exactly three complete cycles so the period would be $\frac{\pi}{3}$.

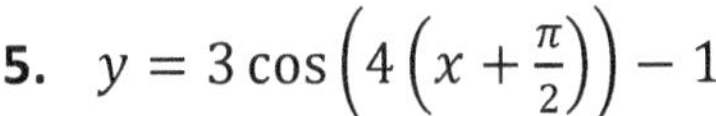

5. $y = 3\cos\left(4\left(x + \frac{\pi}{2}\right)\right) - 1$

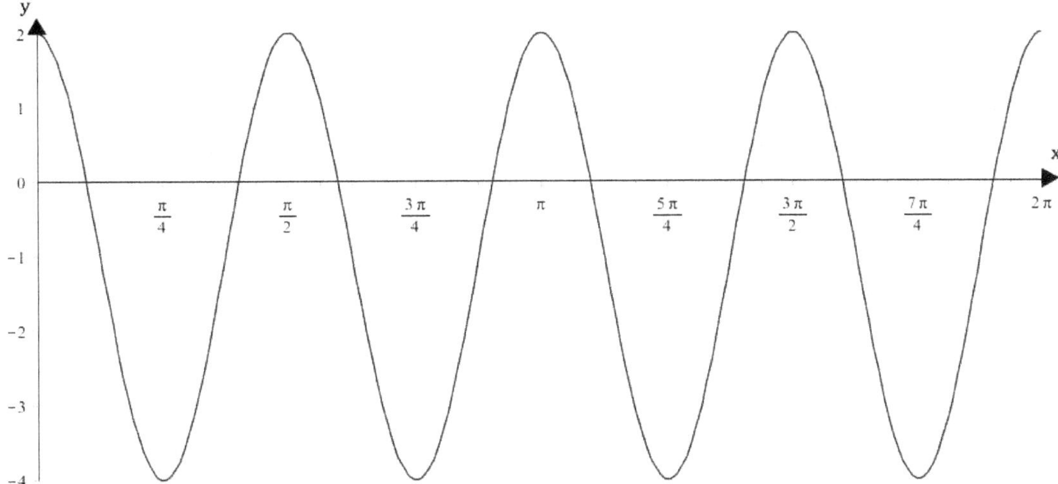

Figure 5.13.

Solution:

$\frac{\pi}{2}$, As you can see, the full cycle is $\frac{\pi}{2}$. That is, the distance between the two initial peaks would be $\frac{\pi}{2}$.

---

6. $y = \cos\left(\frac{(x-3\pi)}{2}\right)$

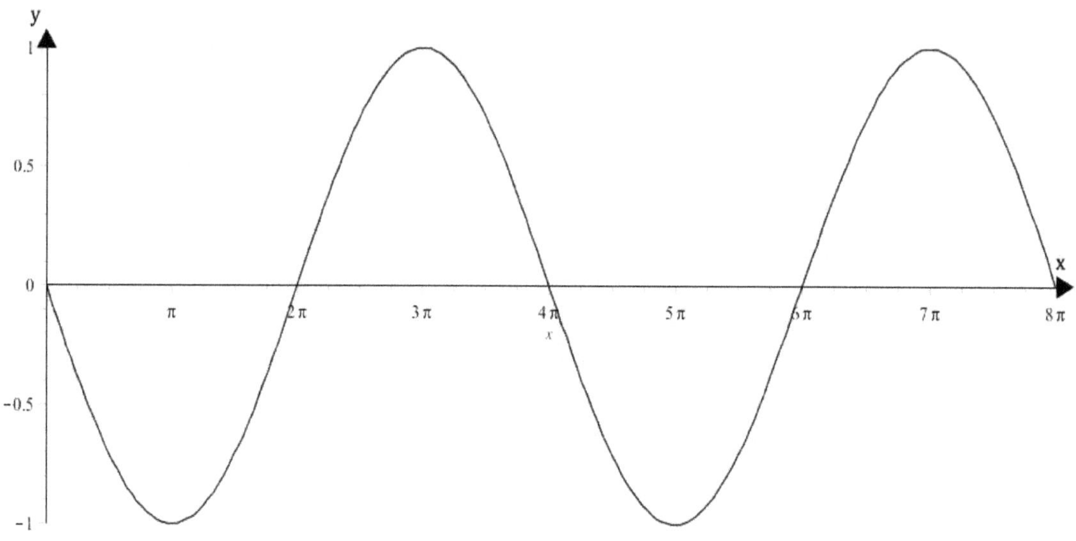

Figure 5.14.

Solution:

4π

5.5. Transformation of sin(x) Function

There is a general formula that we can define as sin(x) or cos(x) function. We start with sin(x):

$$y = A\sin\big(b(x - c)\big) + d$$

In the presented formula A, b, c and d can be defined as follows:

"A" would be amplitude, and it is the distance from the median line to the peak (highest point) or distance from the median line to the trough. In simpler words, it is the amount of vertical compression or stretch. That is, if A is 2, then you would stretch it twice vertically.

"c" is the phase shift, horizontal shift or horizontal translation. It is how much you would move back and forth. Please note the negative sign behind c. That is, if c is positive, you will get a negative number in the formula, and, if c were negative, you would get a positive number. If c is positive, you have to move the function to the right, and, if negative, you have to shift to the left.

"d" is the vertical shift or vertical translation. It states how much you should move the entire graph up or down. Another analogy would be how much you should move the median line up or down. For example, if d is 4, the entire graph would be moved up by 4 units.

"b" can tell us about the period, but it is NOT the period. To find the period, we use b in the following formula:

$$\frac{2\pi}{b} = period\ or\ \frac{2\pi}{period} = b$$

One complete cycle. That is, you start from a certain point (y-value) and a certain direction, and you get back to the same point while keeping the same direction.

Now let's try a few questions about transformation of the sin(x) function, given we know how to graph y = sin(x). Please refer to the following diagram for clarification:

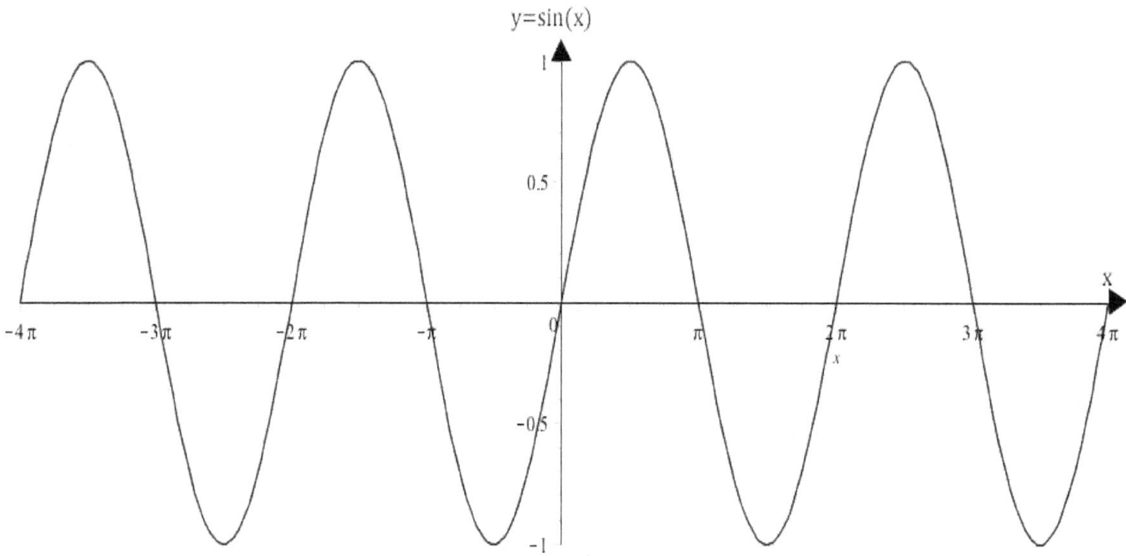

Figure 5.15.

I believe it is better if we graph the functions in the suggested stepwise manner; however, you may notice there are other possibilities to get to the right answer. You can find the list of steps that I found easy to follow.

1. First, try to put the function in the format that we have presented, which is: $y = A\sin\big(b(x-c)\big) + d$.
2. Start with graphing the y = sin(x) function.
3. Always start inside out, that is, first deal with "c", the phase shift (move the sin(x) function to the left or right).
4. Then you take care of "b" to find the period. The period would let us know the length of our function in the x-direction for one complete cycle.
5. The next would be considering amplitude. That is how much you would go up and down finding your max and means in the vertical direction.
6. The vertical translation is the last step that I recommend. Please note you can do it in many other fashions, and you may still get the correct answer. Use your judgment to find your way to draw.

We now use the steps to draw the graph of the function bellow.

$$y = 5\sin(2x - \pi) - 1$$

1. The first step is to put the function in the standard form. The only thing that we need to do is factoring 2, in which case we get:

$$y = 5\sin\left(2(x - \frac{\pi}{2})\right) - 1$$

2. We then graph $y = \sin(x)$:

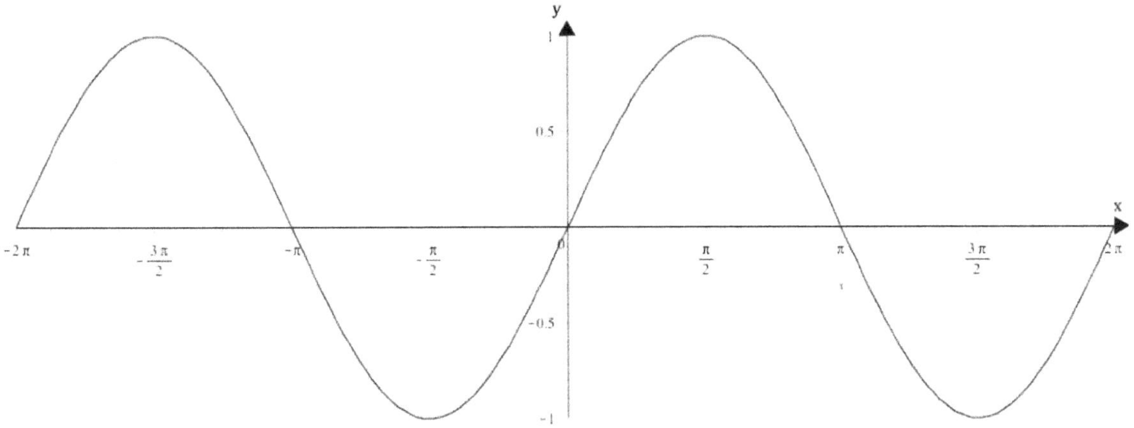

Figure 5.16.

3. To graph $y = sin(x - \frac{\pi}{2})$ we need to move the sin(x) to the right by $\frac{\pi}{2}$ unit. Notice c is $\frac{\pi}{2}$, which we interpret as moving the function $\frac{\pi}{2}$ to the right.

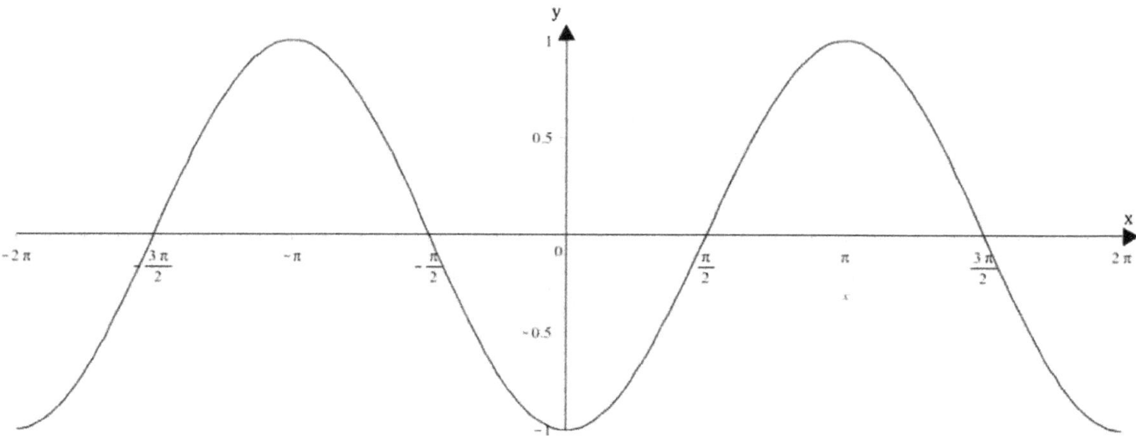

Figure 5.17.

4. To graph $y = \sin\left(2(x - \frac{\pi}{2})\right)$, we have two paths that we can take: first by squeezing the function by two since "b" is 2. The other way to approach it is to find the period by saying b=2. We use the following formula to obtain the period:

$$\frac{2\pi}{b} = period => \frac{2\pi}{2} = \pi$$

Then we would get the following graph:

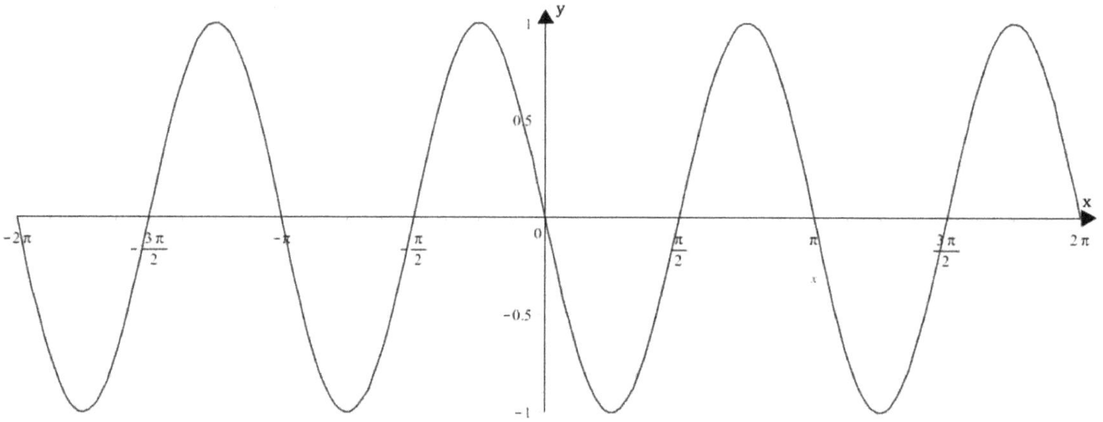

Figure 5.18.

5. The next step would be to graph $y = 5\sin\left(2(x - \frac{\pi}{2})\right)$. With the knowledge that the amplitude "A" is equal to 5, we know we have to stretch the function by five from both top and bottom. That is, the new peak would be at five instead of 1, and the trough would fall on -5 in oppose to -1.

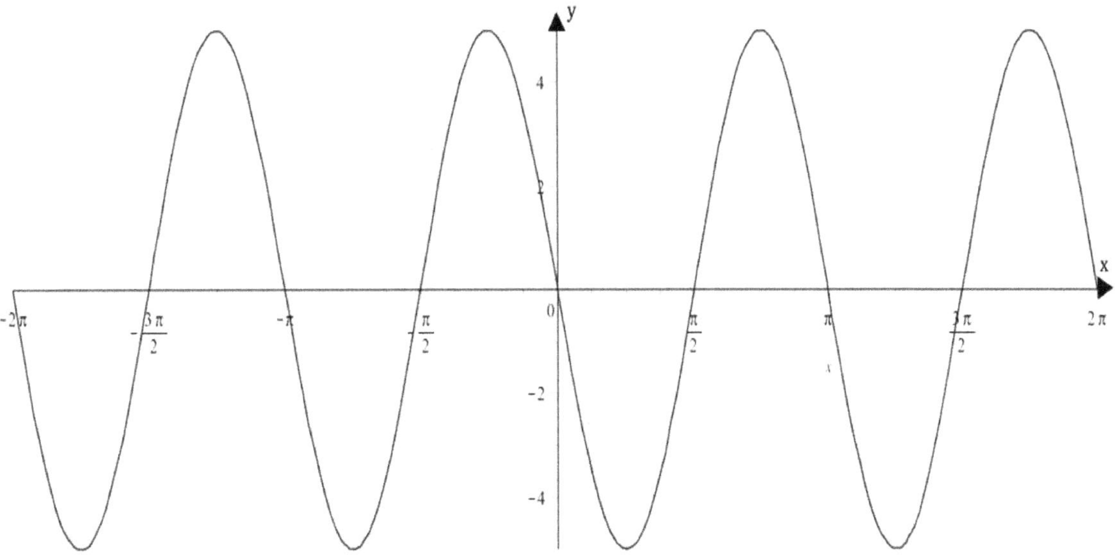

Figure 5.19.

6. The final step is to graph $y = 5\ (\sin\left(2(x - \frac{\pi}{2})\right) - 1$. Knowing "d" is -1, it means we need to move the function one unit down. That is, the new median line would be at y = -1 instead of y = 0 (the x-axis).

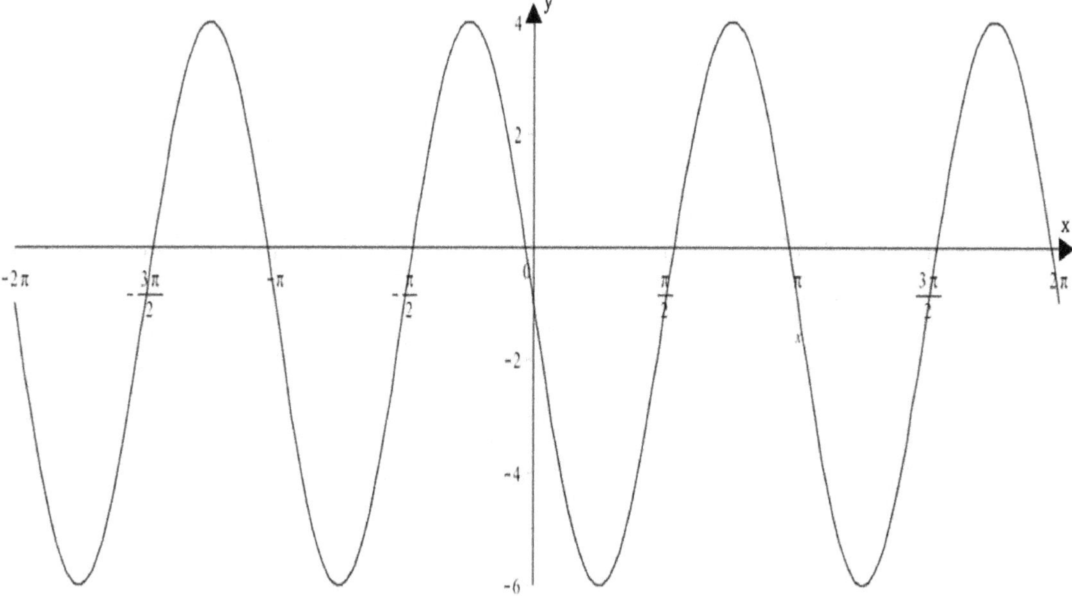

Figure 5.20.

Examples: For each of the following sinusoidal functions, first try to find the amplitude, period, phase shift and vertical translation, and then graph the associated function for values of x between -2π *and* 2π (*or* $x \in [-2\pi, 2\pi]$). Please note you may need to put the function in the standard form first, which is $y = A\sin(b(x - c)) + d$.

7. $y = \sin(x + \frac{\pi}{2})$

Solution:

Amplitude: 1

Period: 2π (b=1 and the period is $\frac{2\pi}{b} = \frac{2\pi}{1}$=$2\pi$)

Phase Shift: $-\frac{\pi}{2}$ (so we have to move the function to the left by $\frac{\pi}{2}$)

Vertical Translation: 0 (so we will not move the function up or down)

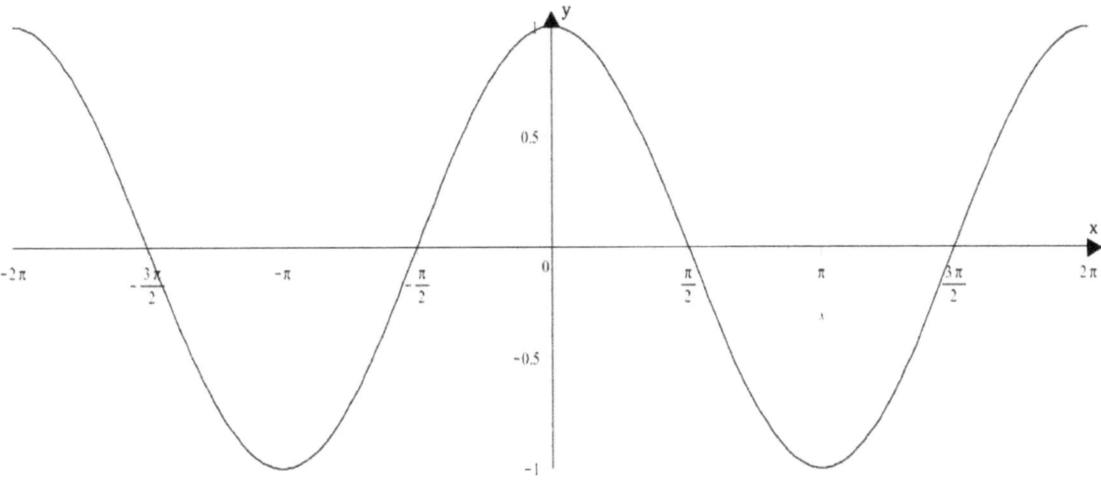

Figure 5.21.

8. $y = \sin\ (x + 1)$

Solution:

Amplitude: 1

Period: 2π (b=1 and the period is $\frac{2\pi}{b} = \frac{2\pi}{1}$=$2\pi$)

Phase Shift: -1 (so we have to move the function to the left by 1)

Vertical Translation: 0 (so we will not move the function up or down)

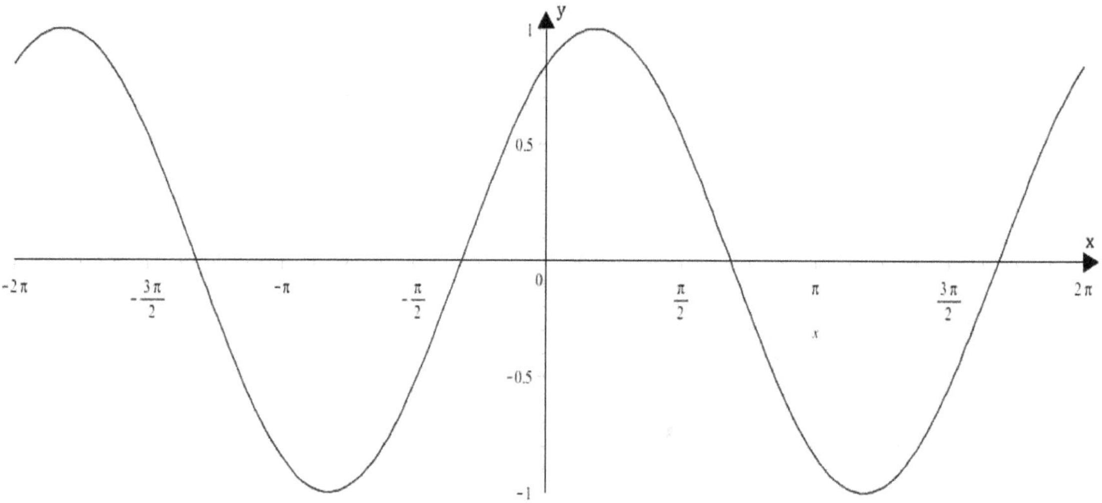

Figure 5.22.

---

9. $y = \sin(x) + 1$

Solution:

Amplitude: 1

Period: 2π (b=1 and the period is $\frac{2\pi}{b} = \frac{2\pi}{1}$=$2\pi$)

Phase Shift: 0 (so we will not move the function left or right)

Vertical Translation: 1 (so we have to move the function up by 1)

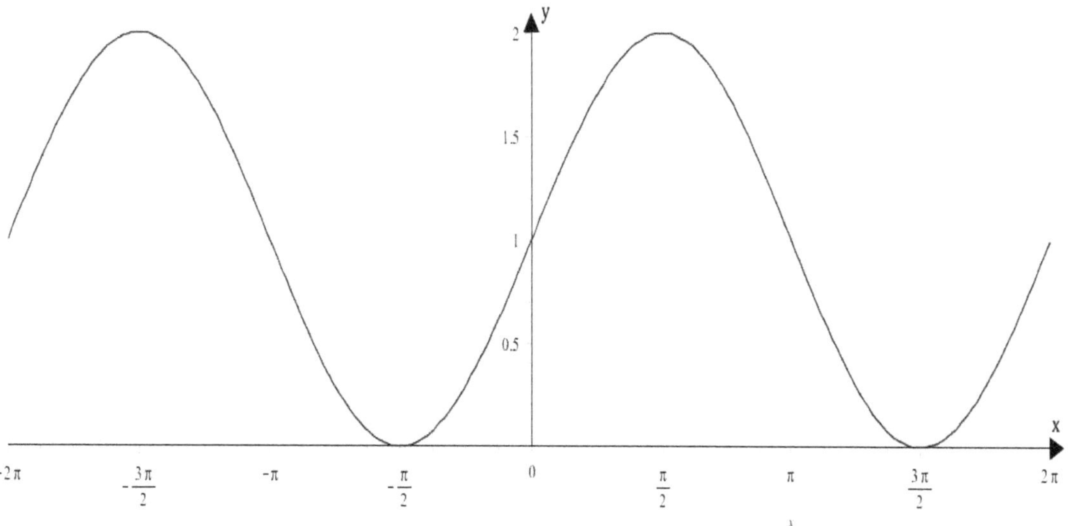

Figure 5.22

---

10. $y = 2\sin(x) + 1$

Solution:

Amplitude: 2

Period: 2π (b=1 and the period is $\frac{2\pi}{b} = \frac{2\pi}{1}$=$2\pi$)

Phase Shift: 0 (so we will not move the function left or right)

Vertical Translation: 1 (so we have to move the function up by 1)

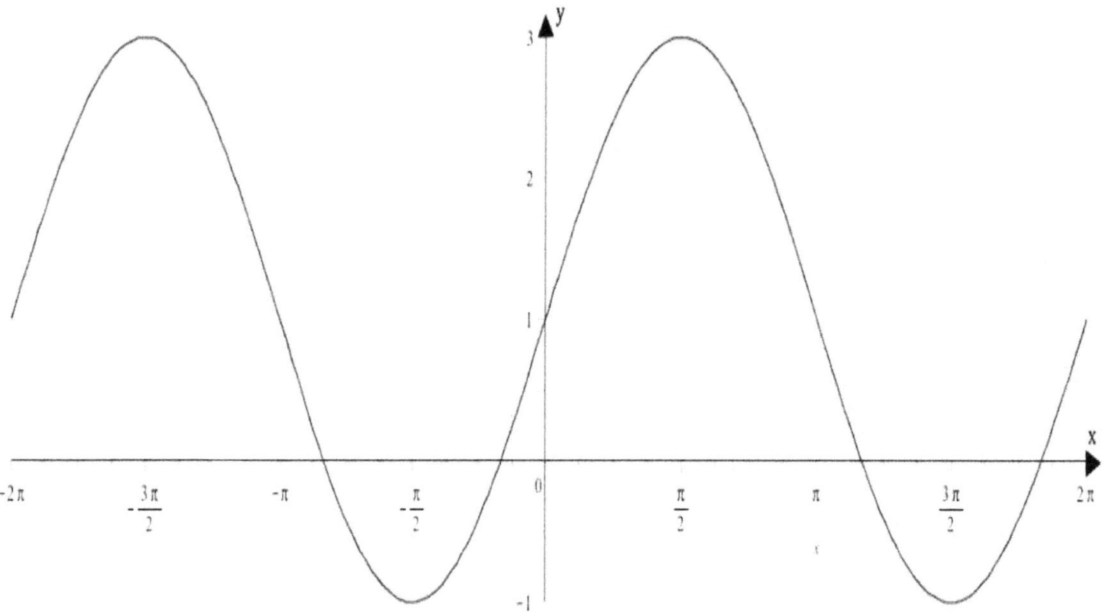

Figure 5.23.

---

11. $y = \sin(2(x-1))$

Solution:

Amplitude: 1

Period: π (b=2 and the period is $\frac{2\pi}{b} = \frac{2\pi}{2} = \pi$)

Phase Shift: 1 (so we have to move the function to the right by 1)

Vertical Translation: 0 (so we will not move the function up or down)

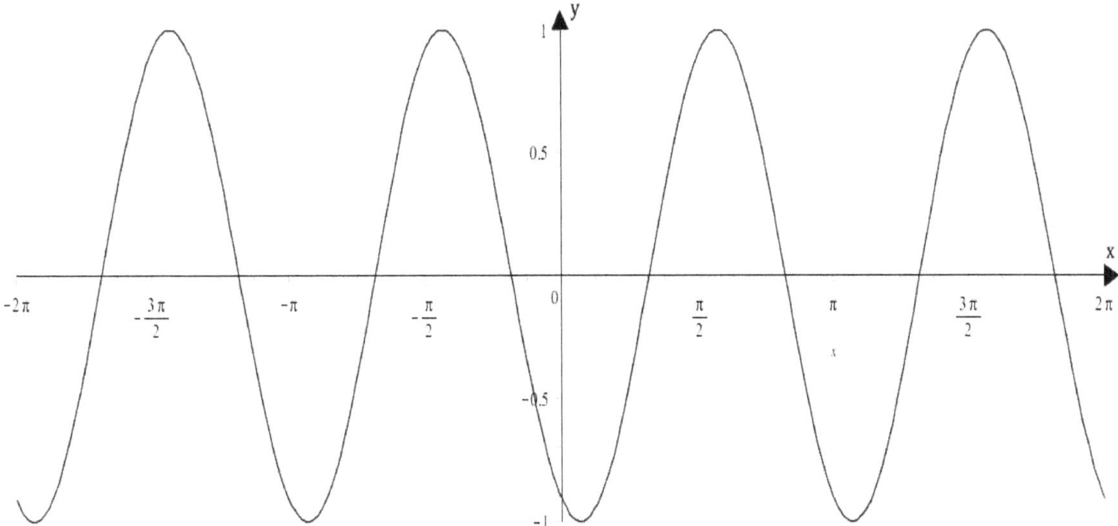

Figure 5.24.

---

12. $y = \frac{\sin(2x)}{2}$

Solution:

For simplicity, I rewrite the function in the form $y = \frac{\sin(2x)}{2} = \frac{1}{2}\sin(2x)$

Amplitude: $\frac{1}{2}$

Period: π (b=2 and the period is $\frac{2\pi}{b} = \frac{2\pi}{2} = \pi$)

Phase Shift: 0 (so we will not move the function left or right)

Vertical Translation: 0 (so we will not move the function up or down)

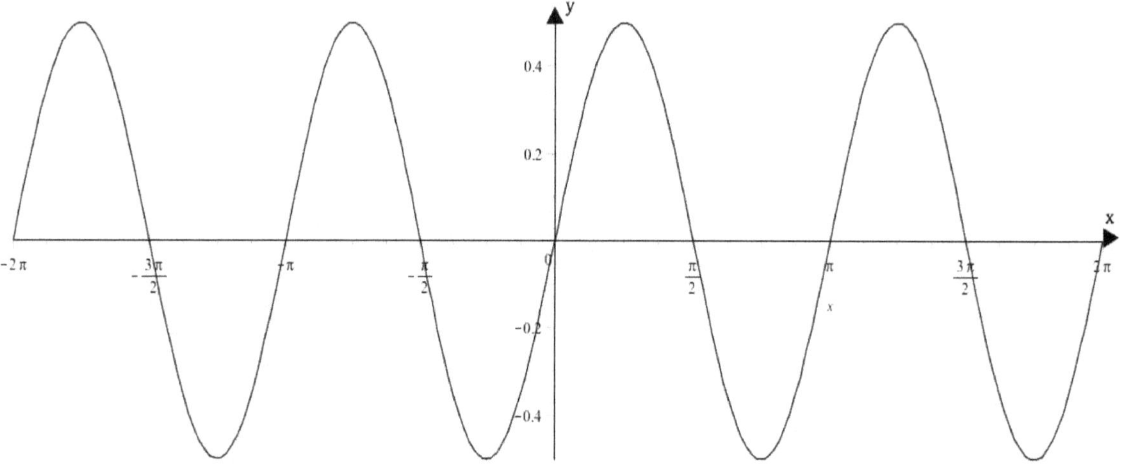

Figure 5.25.

---

13. $y = 2\sin(2x - \pi) + 3$

Solution:

First, we need to rewrite the function in standard form, which is:

$y = A\sin\,(b(x-c)) + d$, so the function can be written as follows:

$y = 2\sin(2x - \pi) + 3 = 2\sin\left(2(x - \frac{\pi}{2})\right) + 3$

Amplitude: 2

Period: π (b=2 and the period is $\frac{2\pi}{b} = \frac{2\pi}{2} = \pi$)

Phase Shift: $\frac{\pi}{2}$ (so we have to move the function to the right by $\frac{\pi}{2}$)

Vertical Translation: 3 (so we have to move up by 3)

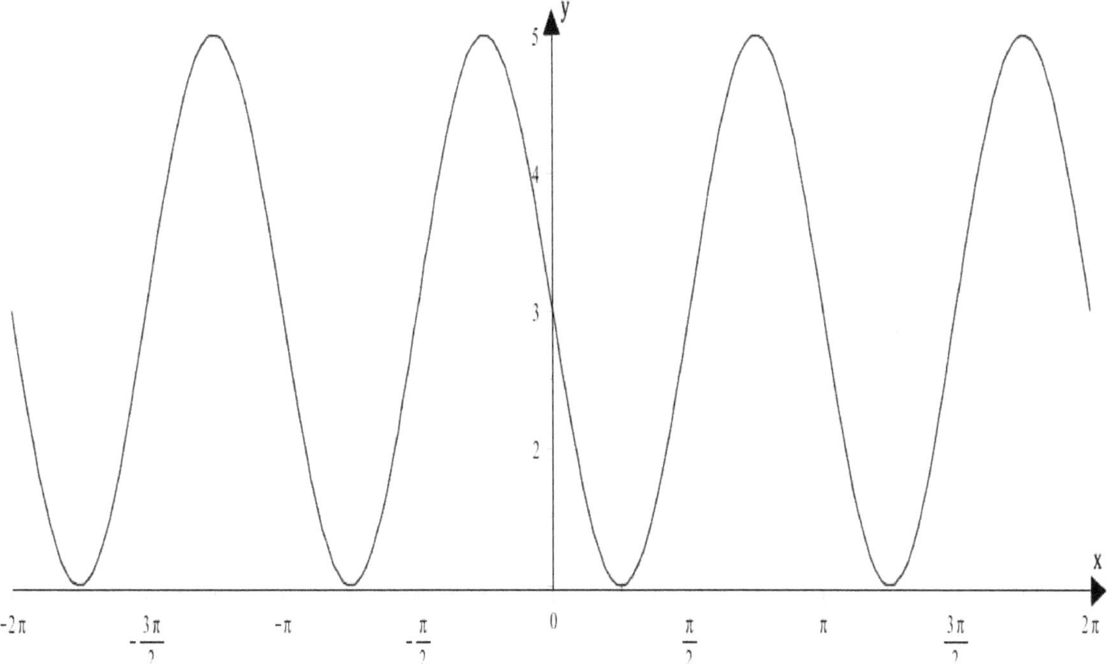

Figure 5.26.

---

14. $y = 3\sin(x - 1) + 1$

Solution:

Amplitude: 3

Period: 2π (b=1 and the period is $\frac{2\pi}{b} = \frac{2\pi}{1}$=$2\pi$)

Phase Shift: 1 (so we have to move right by 1)

Vertical Translation: 1 (so we have to move up by 1)

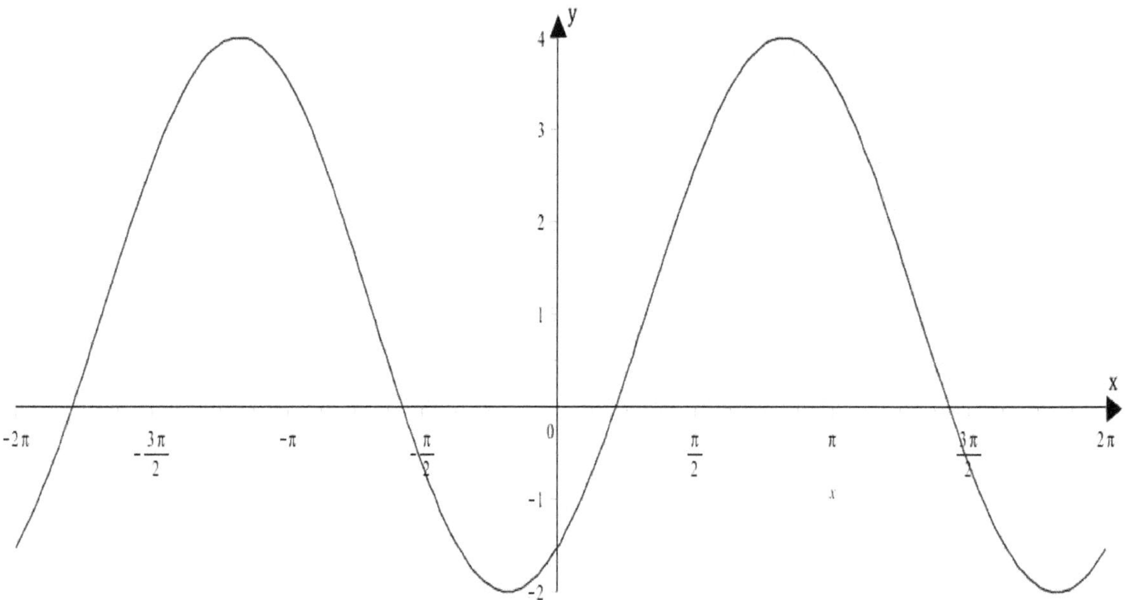

Figure 5.27.

---

15. $y = \frac{2}{5}\sin(x+1) - 3$

Solution:

Amplitude: $\frac{2}{5}$

Period: 2π (b=1 and the period is $\frac{2\pi}{b} = \frac{2\pi}{1}$=$2\pi$)

Phase Shift: -1 (so we have to move the function to the left by 1)

Vertical Translation: -3 (so we have to move the function up by -3)

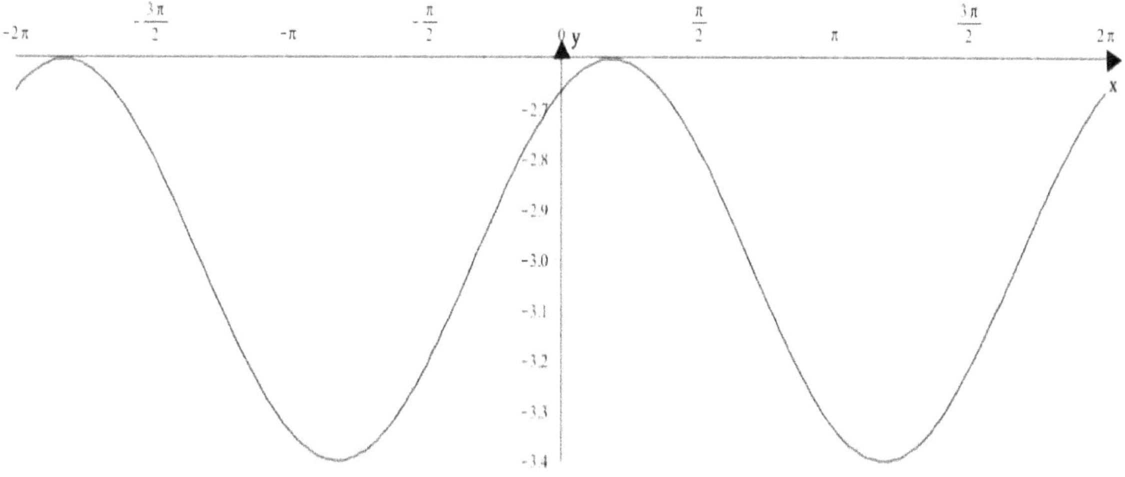

Figure 5.28.

---

16. $y = 3\sin\left(2(x - \frac{\pi}{4})\right) + \frac{1}{2}$

Solution:

Amplitude: 3

Period: π (b=2 and the period is $\frac{2\pi}{b} = \frac{2\pi}{2} = \pi$)

Phase Shift: $\frac{\pi}{4}$ (so we have to move the function to the right by $\frac{\pi}{4}$)

Vertical Translation: $\frac{1}{2}$ (so we have to move the function up by $\frac{1}{2}$)

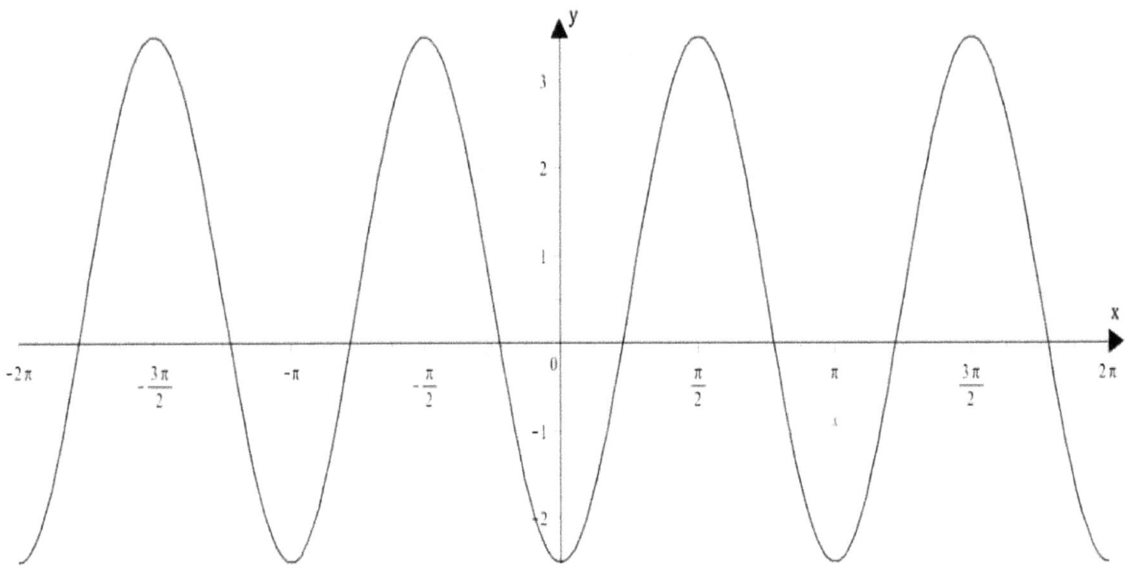

Figure 5.29.

5.6. Transformation of cos(x) Function

Now let's try to define a general formula for cos(x):

$$y = A\cos\big(b(x - c)\big) + d$$

In the presented formula A, b, c and d can be defined as follows:

"A" would be amplitude, and it is the distance from the median line to the peak (highest point) or distance from the median line to trough. In simpler words, it is the amount of vertical compression or stretch. That is, if A is 2, then you would stretch it twice vertically.

"c" is the phase shift, horizontal shift or horizontal translation. It is how much you would move back and forth. Please note the negative sign behind c. That is, if c is positive, you will get a negative number in the formula, and if c were negative, you would get a positive number. If c is positive, you have to move the function to the right, and, if negative, you have to shift to the left.

"d" is the vertical shift or vertical translation. It states how much you should move the entire graph up or down. Another analogy would be how much you should move the median line up or down. For example, if d is 4, the entire graph would be moved up by 4 units.

"b" can tell us about the period, but it is NOT the period. To find the period, we use b in the following formula:

$$\frac{2\pi}{b} = period \ or \ \frac{2\pi}{period} = b$$

One complete cycle. That is, you start from a certain point (y-value) and a certain direction, and you get back to the same point while keeping the same direction.

Now let's try a few questions about transformation of cos(x) function, given that we know how to graph y = cos (x). Please refer to the following diagram for clarification:

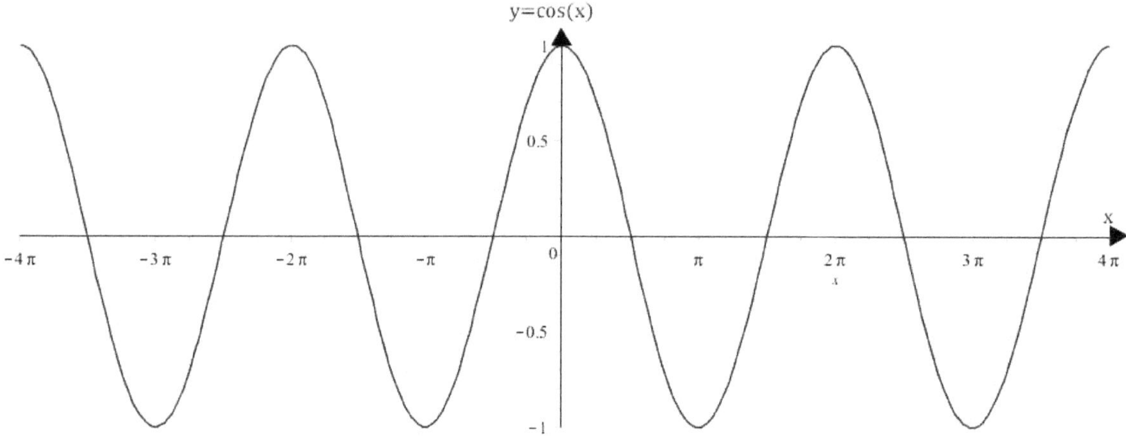

Figure 5.30.

I believe it is better if we graph the functions in the suggested stepwise manner; however, you may notice there are other possibilities to get to the right answer. You can find the list of steps that I found easy to follow.

1. First, try to put the function in the format that we have presented, which is: $y = A\cos\big(b(x-c)\big) + d$.
2. Start with graphing the y = cos(x) function.
3. Always start inside out, that is first deal with "c", the phase shift (move the cos(x) function to the left or right).
4. Then you take care of "b" to find the period. The period would let us know the length of our function in the x-direction for one complete cycle.
5. The next would be considering amplitude. That is how much you would go up and down, finding your max and means in the vertical direction.
6. The vertical translation is the last step that I recommend. Please note you can do it in many other fashions, and you may still get the correct answer. Use your judgment to find your way to draw.

We now use the steps to draw the graph of the function below.

$$y = 2\cos(3x - 2\pi) - 1$$

1. The first step is to put the function in the standard form. The only thing that we need to do is factoring 3, in which case we get:

$$y = 2\cos\left(3\left(x - \frac{2\pi}{3}\right)\right) - 1$$

2. We then graph $y = \cos(x)$

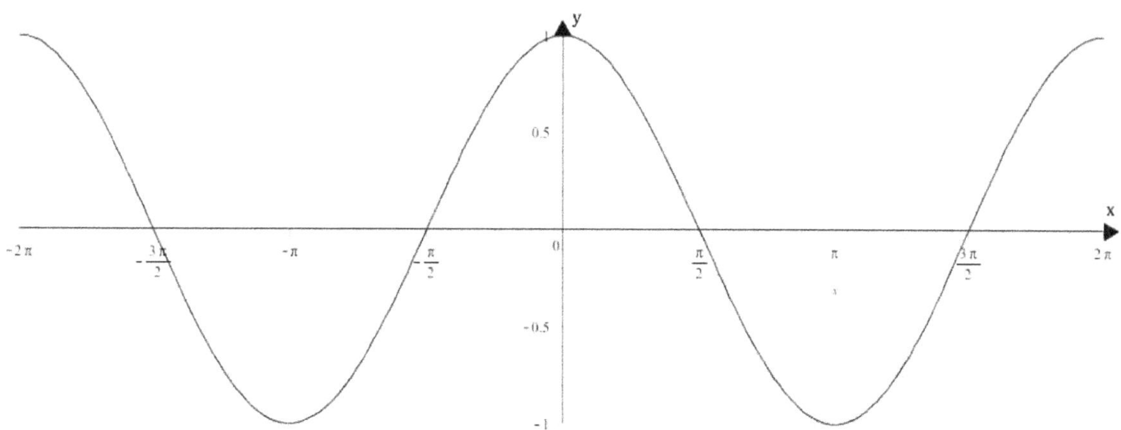

Figure 5.31.

3. To graph $y = cos\left(x - 2\frac{\pi}{3}\right)$, we need to move the cos(x) to the right by $\frac{2\pi}{3}$ unit. Notice c is $\frac{2\pi}{3}$, which we interpret as moving the function $\frac{2\pi}{3}$ to the right.

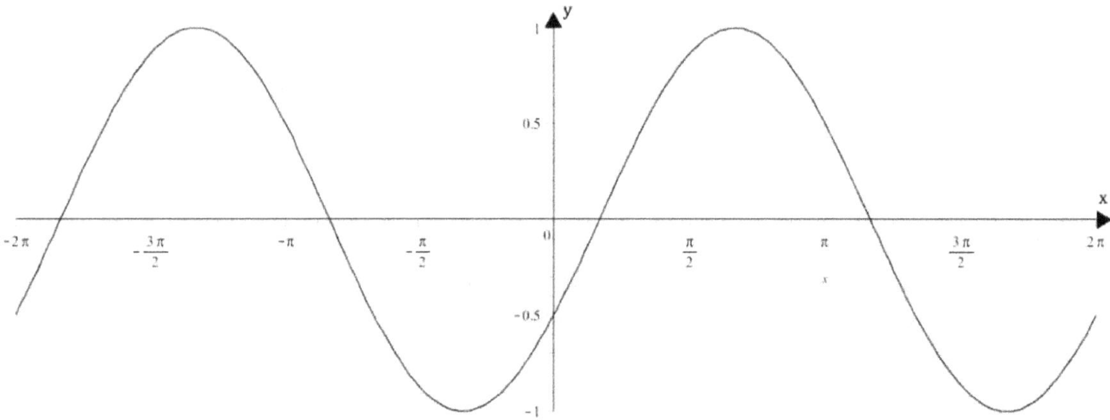

Figure 5.32

4. To graph $y = cos\left(3\left(x - 2\frac{\pi}{3}\right)\right)$, we have two paths that we can take, first by squeezing the function by three since "b" is 3. The other way to approach is to find the period by saying b=3. We use the following formula to obtain the period:

$$\frac{2\pi}{b} = period => \frac{2\pi}{3}$$

Then we would get the following graph:

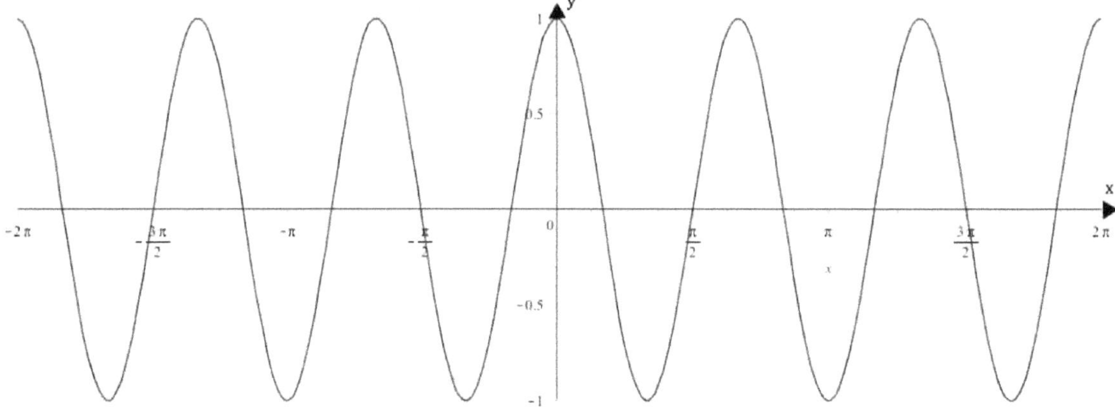

Figure 5.33.

5. The next step would be to graph $y = 2\ cos\left(3\left(x - 2\frac{\pi}{3}\right)\right)$. With the knowledge that the amplitude "A" is equal to 2, we know we have to stretch the function by two from both top and bottom. That is, the new peak would be at two instead of 1, and the trough would fall on -2 as opposed to -1.

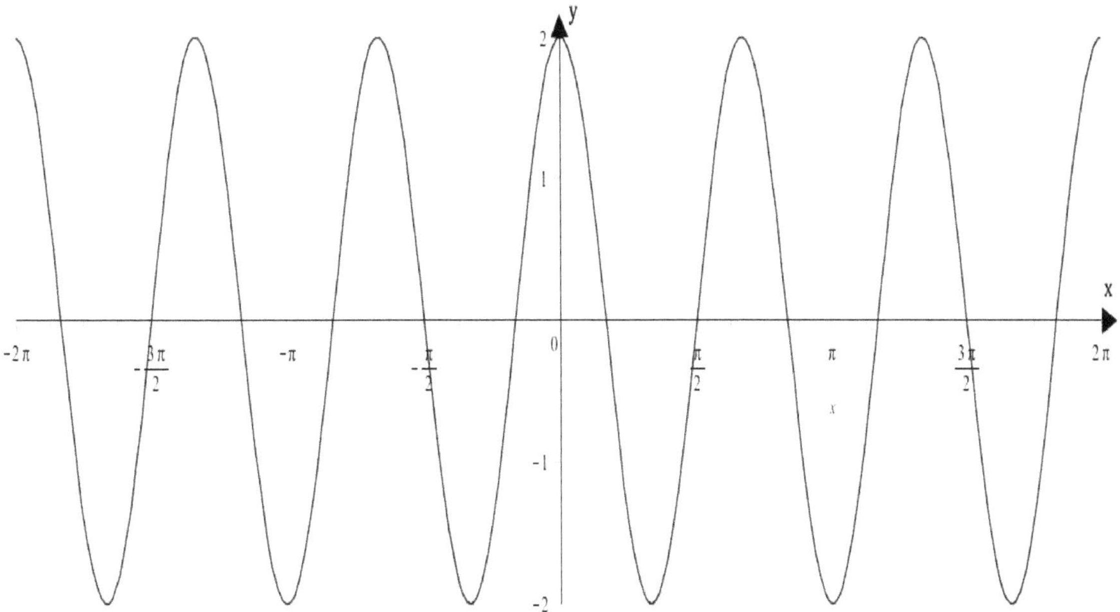

Figure 5.34.

6. The final step is to graph $y = 2(\cos\left(3(x - \frac{2\pi}{3})\right) - 1$. Knowing "d" is -1, it means we need to move the function one unit down. That is, the new median line would be at y = -1 instead of y = 0 (the x-axis).

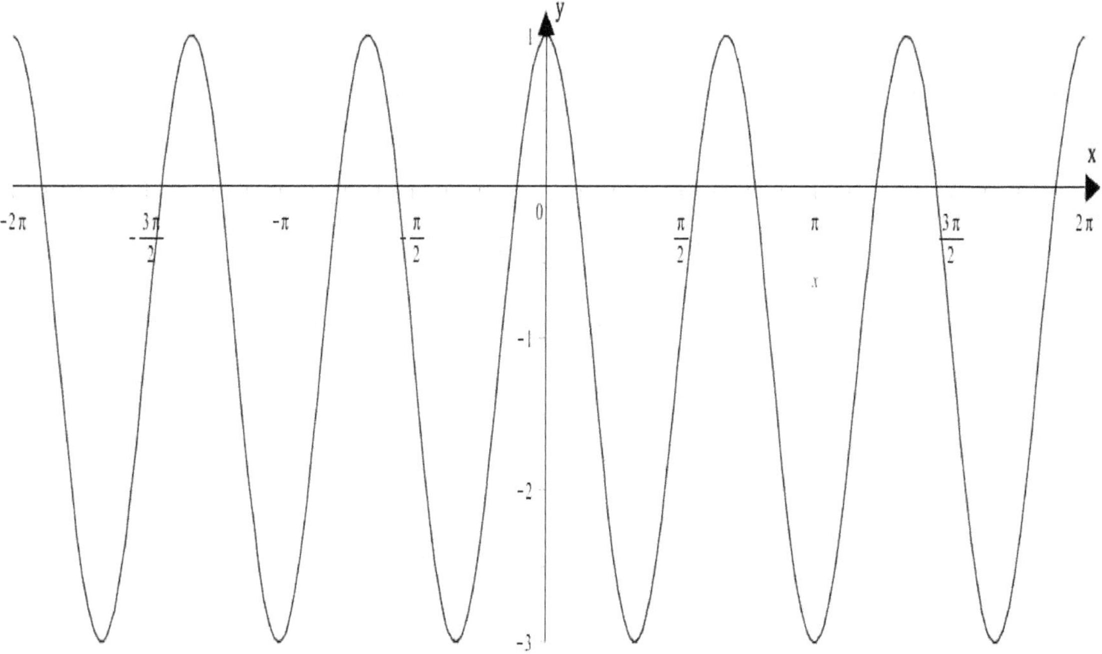

Figure 5.35.

Examples: For each of the following sinusoidal functions, first try to find the amplitude, period, phase shift and vertical translation, and then graph the associated function for values of x between -2π *and* 2π (*or* $x \in [-2\pi, 2\pi]$). Please note you may need to put the function in the standard form first, which is $y = A\cos(b(x-c)) + d$.

7. $y = \cos(2x)$

Solution:

Amplitude: 1

Period: π (b=2 and period is $\frac{2\pi}{b} = \frac{2\pi}{2} = \pi$)

Phase Shift: 0 (so we do not have to move the function to the left or right)

Vertical Translation: 0 (so we do not have to move up or down)

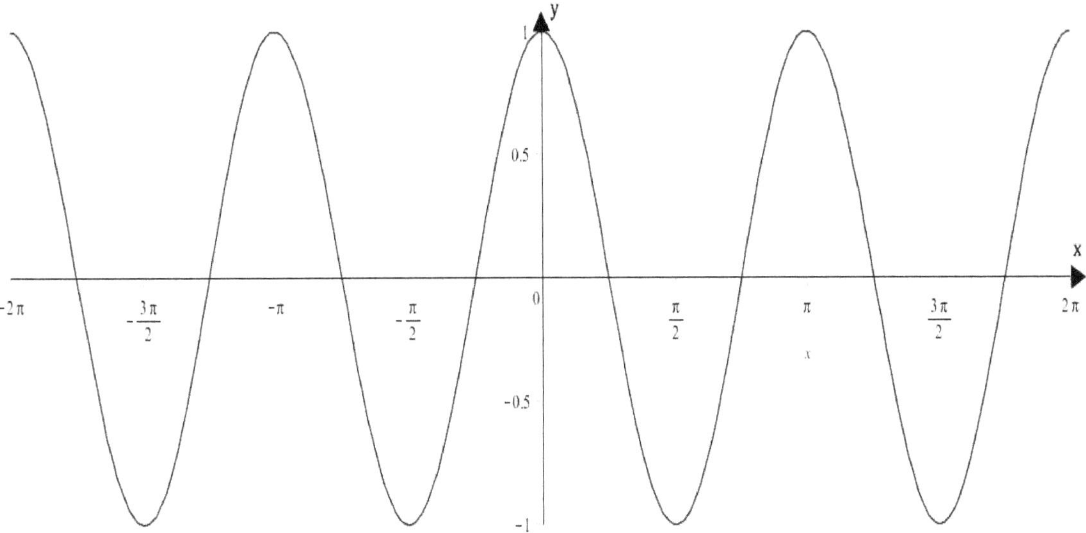

Figure 5.36.

---

8. $y = \cos\left(x + \frac{\pi}{2}\right)$

Solution:

Amplitude: 1

Period: 2π (b=1 and the period is $\frac{2\pi}{b} = \frac{2\pi}{1} = 2\pi$)

Phase Shift: $-\frac{\pi}{2}$ (so we have to move the function to the left by $\frac{\pi}{2}$)

Vertical Translation: 0 (so we do not have to move the function up or down)

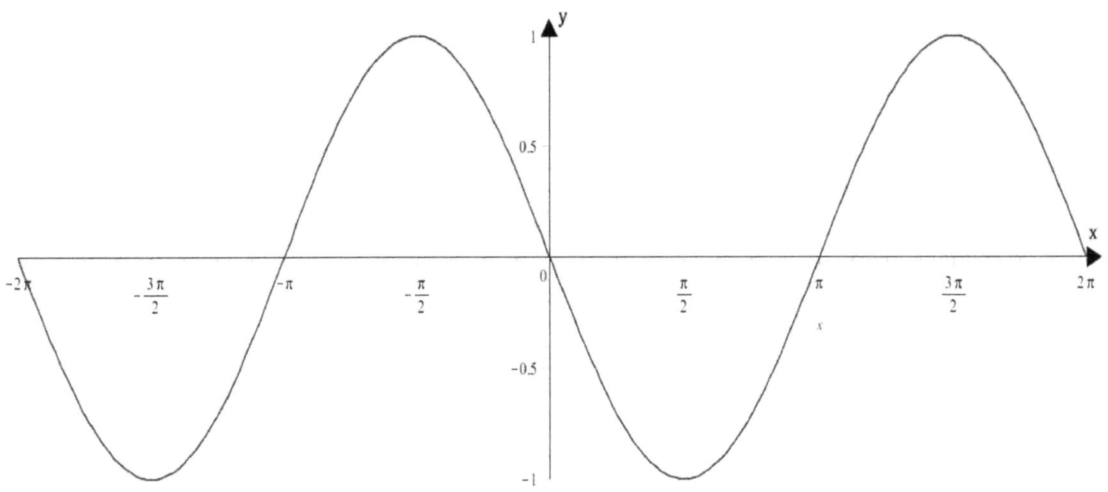

Figure 5.37.

---

9. $y = \cos(x) + 2$

Solution:

Amplitude: 1

Period: 2π (b=1 and the period is $\frac{2\pi}{b} = \frac{2\pi}{1}$=$2\pi$)

Phase Shift: 0 (so we do not have to move the function to the left or right)

Vertical Translation: 2 (so we have to move the function up by 2)

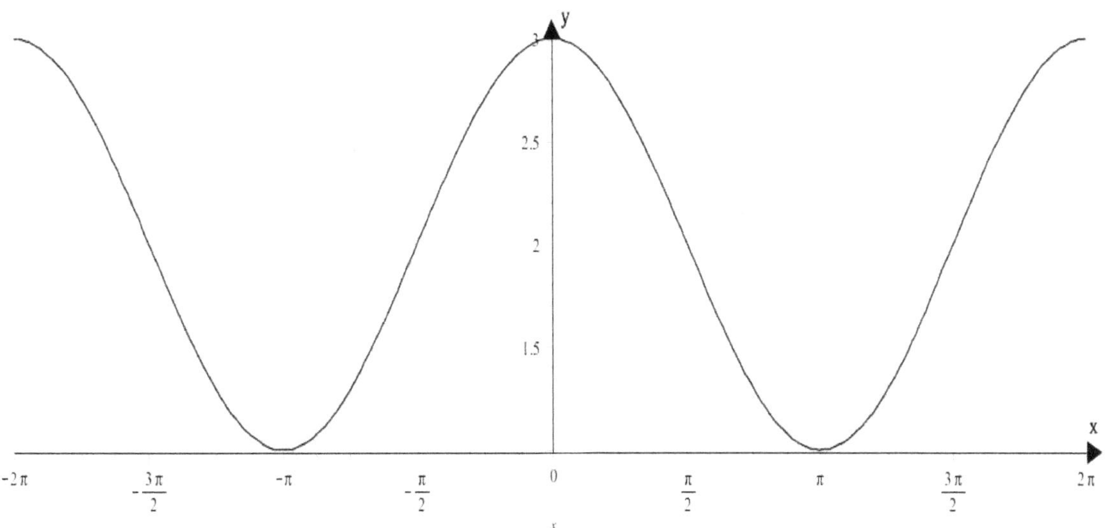

Figure 5.38.

---

10. $y = \cos(\frac{x}{2})$

Solution:

It may be clearer if we rewrite the function as:

$$y = \cos\left(\frac{x}{2}\right) = \cos(\frac{1}{2}x)$$

Amplitude: 1

Period: 4π (b $= \frac{1}{2}$ and the period is $\frac{2\pi}{b} = \frac{2\pi}{\frac{1}{2}} = 2\pi\frac{2}{1} = 4\pi$).

Please note the period is 4π, and we are asked to graph the function for the value of x that lies in the interval $[-2\pi\,,\,2\pi]$. That is, we can only have a cycle in this domain.

Phase Shift: 0 (so we do not have to move the function to the left or right).

Vertical Translation: 0 (so we do not have to move the function up or down).

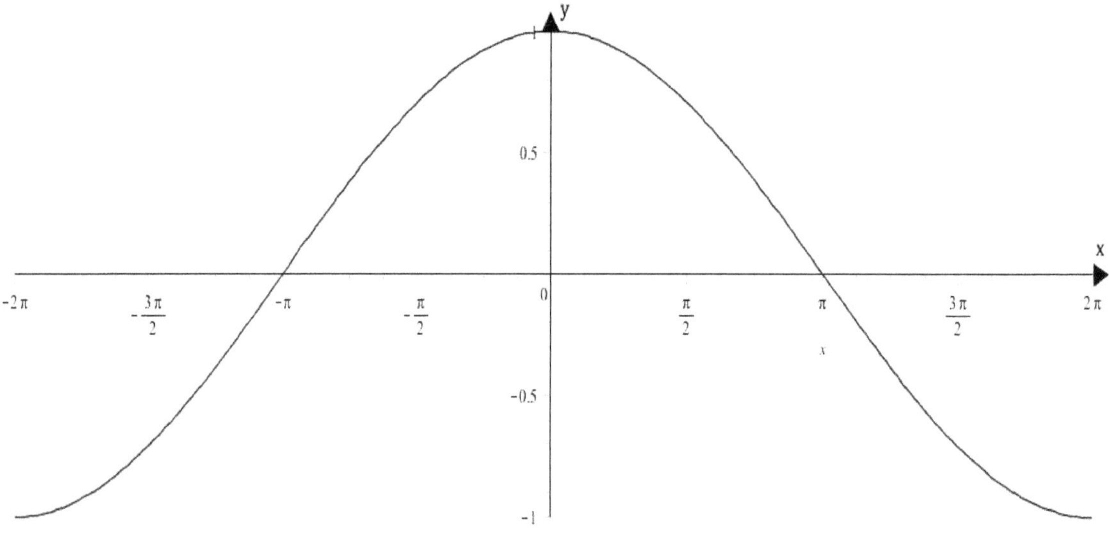

Figure 5.39.

---

11. $\mathrm{y} = \cos\big(4(x-1)\big)$

Solution:

Amplitude: 1

Period: $\frac{\pi}{2}$ (b=4 and the period is $\frac{2\pi}{b} = \frac{2\pi}{4} = \frac{\pi}{2}$)

Phase Shift: 1 (so we have to move the function to the right by 1)

Vertical Translation: 0 (so we do not have to move up or down)

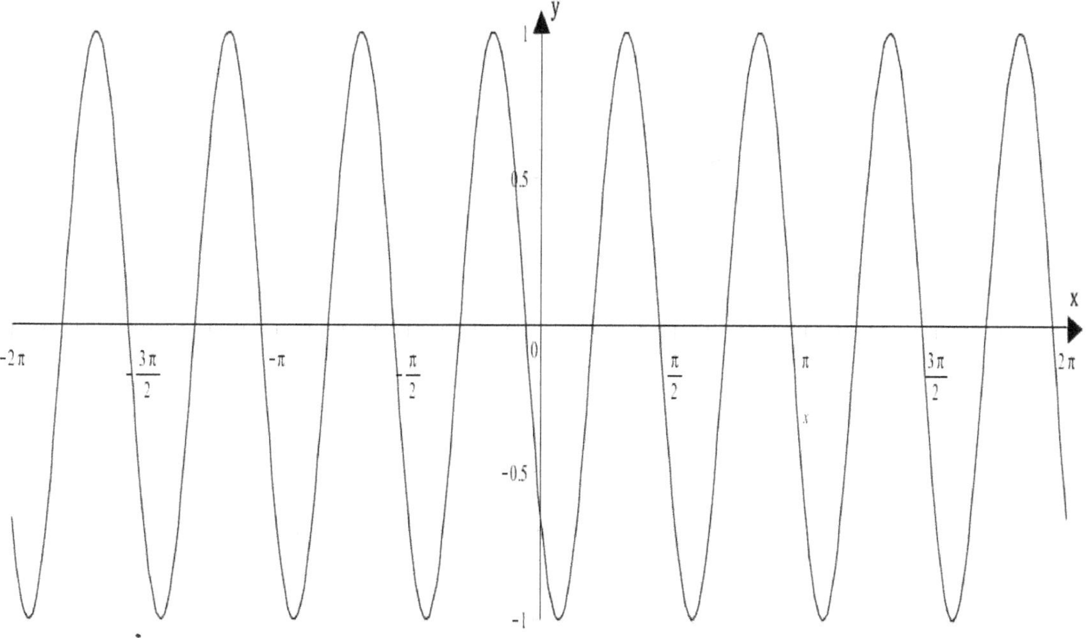

Figure 5.40.

---

12. $y = \cos(3x - \pi)$

Solution:

First we need to rearrange to be in standard form by factoring out 3. This way we get:

$$y = \cos(3x - \pi) = \cos\left(3\left(x - \frac{\pi}{3}\right)\right)$$

Amplitude: 1

Period: $\frac{2\pi}{3}$ (b=3 and the period is $\frac{2\pi}{b} = \frac{2\pi}{3}$)

Phase Shift: $\frac{\pi}{3}$ (so we have to move the function to the right by $\frac{\pi}{3}$)

Vertical Translation: 0 (so we do not have to move the function up or down)

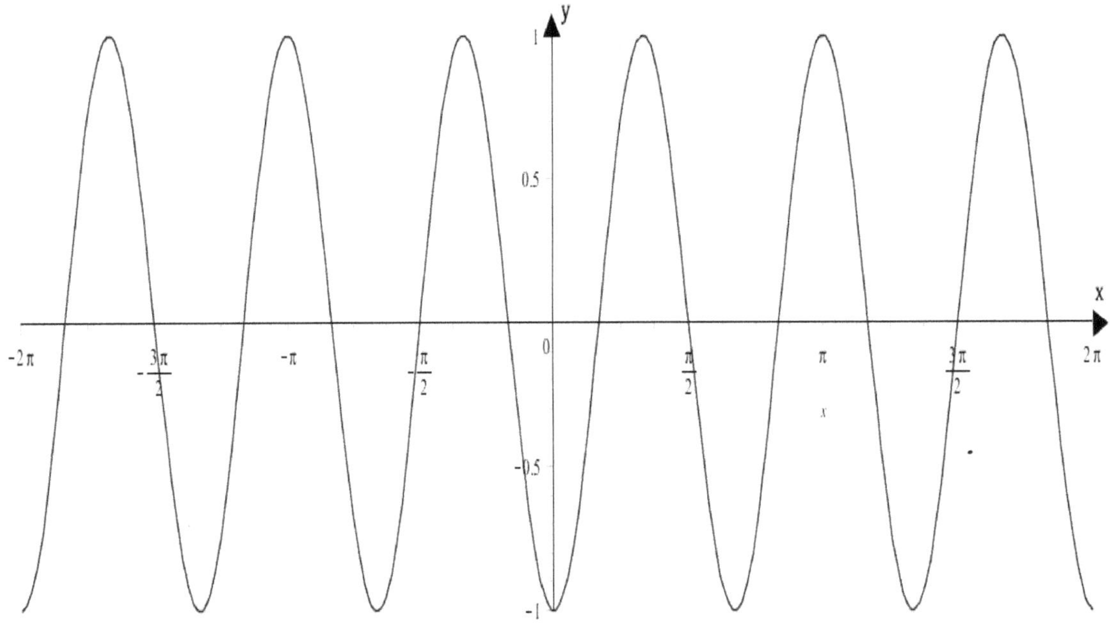

Figure 5.41.

13. $y = 4\cos(x - 1)$

Solution:

Amplitude: 4

Period: 2π(b=1 and period is $\frac{2\pi}{b} = \frac{2\pi}{1} = 2\pi$)

Phase Shift: 1 (so we have to move the function to the right by 1)

Vertical Translation: 0 (so we do not have to move the function up or down)

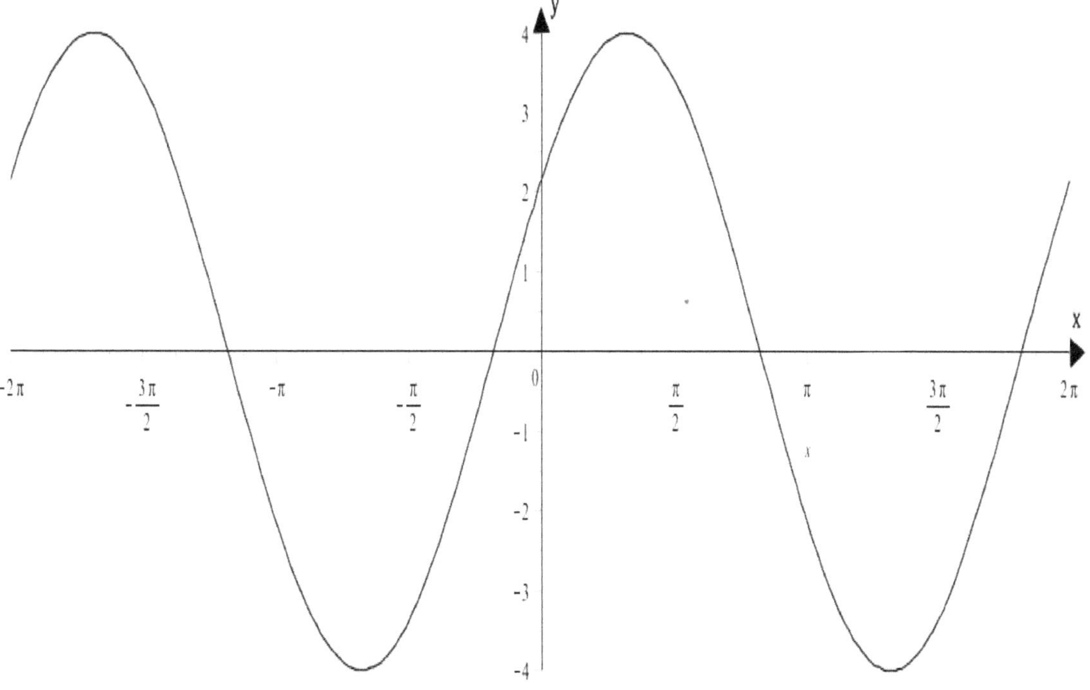

Figure 5.42.

14. $y = \frac{5}{6}\cos\left(x - \frac{\pi}{2}\right) - 1$

Solution:

Amplitude: $\frac{5}{6}$

Period: 2π (b=1 and the period is $\frac{2\pi}{b} = \frac{2\pi}{1} = 2\pi$).

Phase Shift: $\frac{\pi}{2}$ (so we have to move the function to the right by $\frac{\pi}{2}$)

Vertical Translation: -1 (so we have to move the function down by 1).

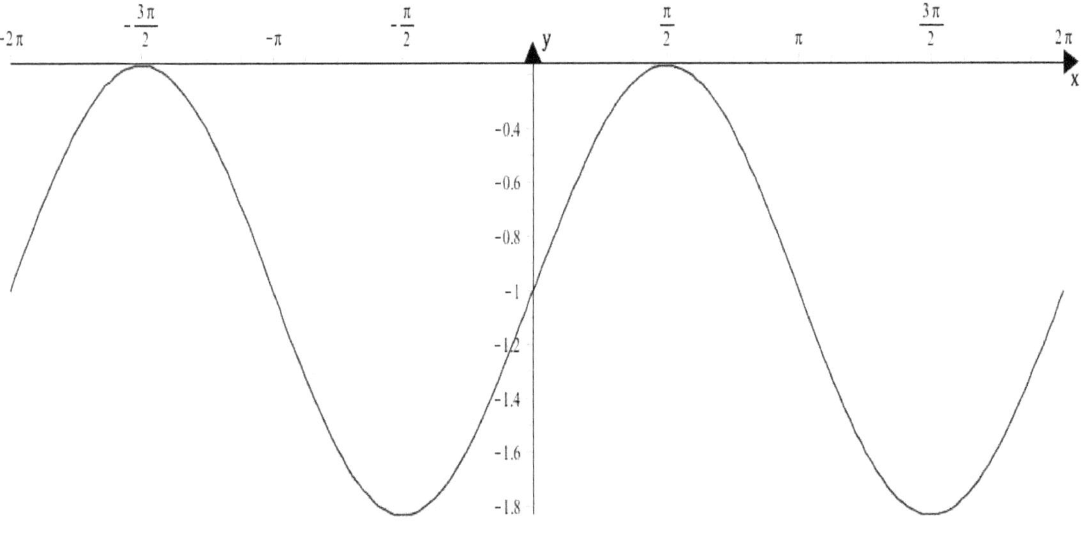

Figure 5.43.

---

15. $y = \frac{\pi}{4}\cos\left(3x - \frac{\pi}{2}\right) + 2$

Solution:

16. First we need to rearrange to be in standard form by factoring out 3. This way we get:

$$y = \frac{\pi}{4}\cos\left(3x - \frac{\pi}{2}\right) + 2 = \frac{\pi}{4}\cos\left(3\left(x - \frac{\pi}{6}\right)\right) + 2$$

Amplitude: $\frac{\pi}{4}$ (remember π is just a number so as $\frac{\pi}{4}$)

Period: $\frac{2\pi}{3}$ (b=1 and the period is $\frac{2\pi}{b} = \frac{2\pi}{3} = \frac{2\pi}{3}$)

Phase Shift: $\frac{\pi}{6}$ (so we have to move the function to the right by $\frac{\pi}{6}$)

Vertical Translation: 2 (so we have to move the function up by 2)

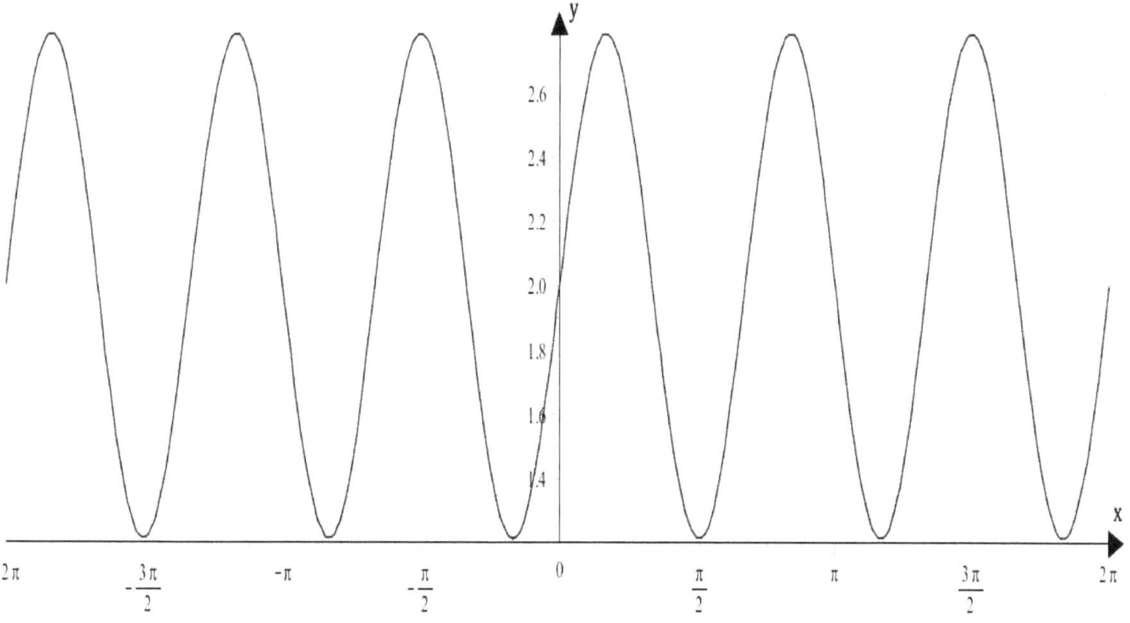

Figure 5.44.

Side Note:

Questions:

Are sin(x) and cos(x) function the same functions only with a small phase shift (left and right translation)?

Absolutely yes!

Then why did we introduce them in two separate parts?

My bad! I could have done it in one. The only reason was I wanted you to practise more and notice this yourself.

Now try to stare at the sin(x) and cos(x) functions to verify the following equality:

$\sin(x) = \cos\left(x - \frac{\pi}{2}\right)$ and $\cos(x) = \sin\left(x + \frac{\pi}{2}\right)$

In other words, if you move sin(x) to the left by $\frac{\pi}{2}$, you would get cos(x). Also, if you shift cos(x) to the right by $\frac{\pi}{2}$, you will obtain sin(x). That is, with a phase shift of $\frac{\pi}{2}$ to the left or right, you can obtain the other function.

To better understand the aforementioned side note, please refer to the following graph and verify the equality $\sin(x) = \cos\left(x - \frac{\pi}{2}\right)$.

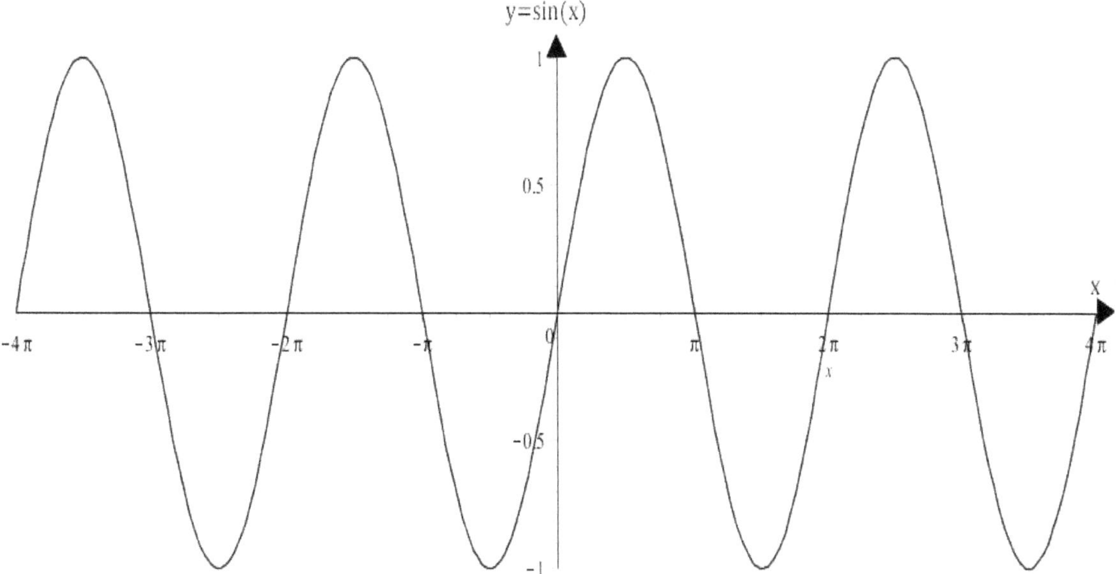

Figure 5.45.

Also, we can see $\cos(x) = \sin\left(x + \frac{\pi}{2}\right)$

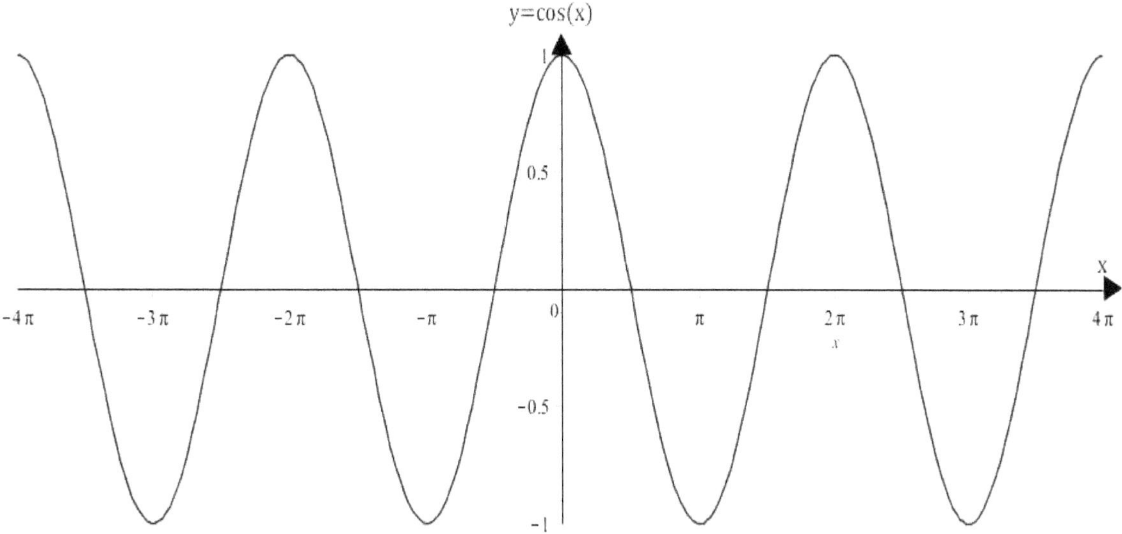

Figure 5.46.

5.7. Simple Way to Graph tan(X), and cot(x)

To graph tan(x) or cot(x), one easy way would be to draw two perpendicular lines (a coordinate system). The vertical line can be your function (for example y = tan(x)), and the horizontal axis can show the magnitude of the associated angles, which are the x values. The unit for the horizontal axis, which are the x values, can be in radian. However, you may try other units of angles, such as degrees.

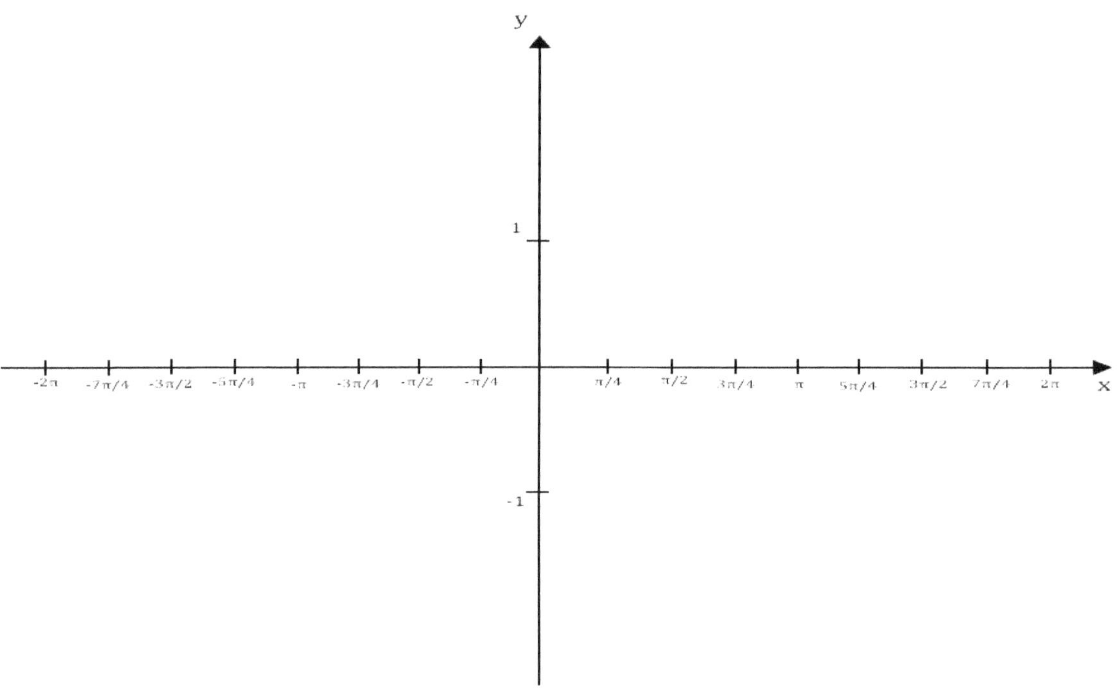

Figure 5.47.

Initially, we are going to graph y=tan(x) in a very simple fashion by finding the coordinates of the point associated to tan(x). To do so, try to find values of tan(x) for a few critical angles. You may find the values of the associated points from the table in chapter one, which we constructed.

$\tan(0) = 0$, which is point → $(0, 0)$

$\tan\left(\frac{\pi}{4}\right) = 1$, which is point → $(\frac{\pi}{4}, 1)$

$\tan\left(\frac{2\pi}{4}\right) = \tan\left(\frac{\pi}{2}\right) \approx \pm\infty$, which is point → $(\frac{\pi}{2}, \pm\infty)$

Why is tan $\left(\frac{\pi}{2}\right) \approx \pm\infty$?

First of all, don't show this to mathematicians, because $\tan\left(\frac{\pi}{2}\right)$ is undefined, and they would throw the book out of the window. Since we have not talked about the limit, I will use the meaning of it without explicitly telling you I am using it (so what am I using? Nothing!). We have defined $\tan(x) = \left(\frac{\sin(x)}{\cos(x)}\right)$ Now let's try to find $\tan\left(\frac{\pi}{2}\right)$= $\left(\frac{\sin(\frac{\pi}{2})}{\cos(\frac{\pi}{2})}\right)$= $\frac{1}{0}$, which is something we cannot solve. However, instead of $\frac{\pi}{2}$, let's get close to $\frac{\pi}{2}$, but not actually get to it. If you use your calculator to find $\tan\left(\frac{\pi}{2} - 0.00000001\right)$ and $\tan\left(\frac{\pi}{2} + 0.00000001\right)$, you will get the following two results. That is, $\tan\left(\frac{\pi}{2} - a\ small\ value\right) =$ 99,999,999.9954 or a great positive number $+\infty$. Moreover, using the calculator, we obtain $\tan\left(\frac{\pi}{2} + a\ small\ value\right) =$ -99,999,999.9954 or a huge negative number symbolized by $-\infty$. Long story short, if you approach point $\frac{\pi}{2}$ from the left (a bit less than $\frac{\pi}{2}$), the value of $\tan\left(\frac{\pi}{2} - a\ small\ value\right)$ would get close to $+\infty$. Similarly, if you approach from the right of $\frac{\pi}{2}$, the value of $\tan\left(\frac{\pi}{2} + a\ small\ value\right)$ would get close to $-\infty$. Remember $\tan\left(\frac{\pi}{2}\right)$ = (tangent exactly at $\frac{\pi}{2}$) is undefined!

$\tan\left(\frac{3\pi}{4}\right)$ =-1 , which is the point → $(\frac{3\pi}{4}, -1)$

$\tan\left(\frac{4\pi}{4}\right) = \tan(\pi) = 0$, which is the point → $(\pi, 0)$

$\tan\left(\frac{5\pi}{4}\right) = 1$, which is the point → $(\frac{5\pi}{4}, 1)$

$\tan\left(\frac{6\pi}{4}\right) = \tan\left(\frac{3\pi}{2}\right) \approx \pm\infty$, which is the point → $(\frac{3\pi}{2}, \pm\infty)$

$\tan\left(\frac{7\pi}{4}\right) = -1$, which is the point → $(\frac{7\pi}{4}, -1)$

$\tan\left(\frac{8\pi}{4}\right) = \tan(2\pi) = 0$, which is the point → $(2\pi, 0)$

$\tan\left(-\frac{\pi}{4}\right) = -1$, which is the point → $(-\frac{\pi}{4}, -1)$

$\tan\left(-\frac{2\pi}{4}\right) = \tan\left(-\frac{\pi}{2}\right) \approx \pm\infty$, which is the point → $(-\frac{\pi}{2}, \pm\infty)$

$\tan\left(-\frac{3\pi}{4}\right) = 1$, which is the point → $(-\frac{3\pi}{4}, 1)$

$\tan\left(-\frac{4\pi}{4}\right) = \tan(-\pi) = 0$, which is the point → $(-\pi, 0)$

$\tan\left(-\frac{5\pi}{4}\right) = -1$, which is the point → $(-\frac{5\pi}{4}, -1)$

$\tan\left(-\frac{6\pi}{4}\right) = \tan\left(-\frac{3\pi}{2}\right) \approx \pm\infty$, which is the point → $(-\frac{3\pi}{2}, \pm\infty)$

$\tan\left(-\frac{7\pi}{4}\right) = 1$, which is the point → $(-\frac{7\pi}{4}, 1)$

$\tan\left(-\frac{8\pi}{4}\right) = \tan(-2\pi) = 0$, which is the point → $(-2\pi, 0)$

Now try to place the obtained points on the coordinate system that you have created.

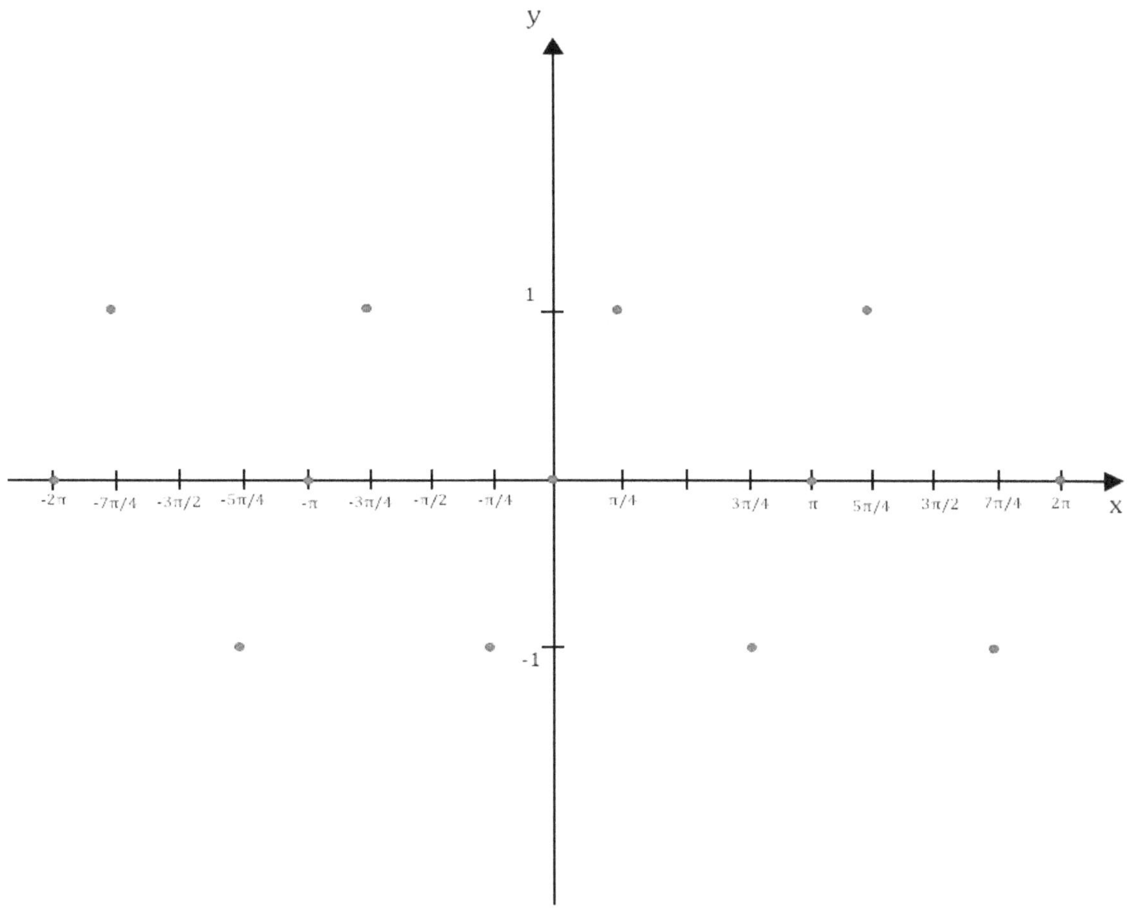

Figure 5.48.

Now let's take care of points such as $\frac{\pi}{2}, \frac{3\pi}{2}, -\frac{\pi}{2}$, and $-\frac{3\pi}{2}$. As you have noticed, the values at the aforementioned points are undefined since we are dividing by zero. Now we show these restricted points with the lines that our function is not supposed to touch. Let's call them Vertical Asymptotes. On the other side, let's show points of infinity with an arrow. The arrows pointing towards the positive values of y (upwards) are representing +∞ while the downwards arrows, which are pointing towards negative y values, are expressing -∞.

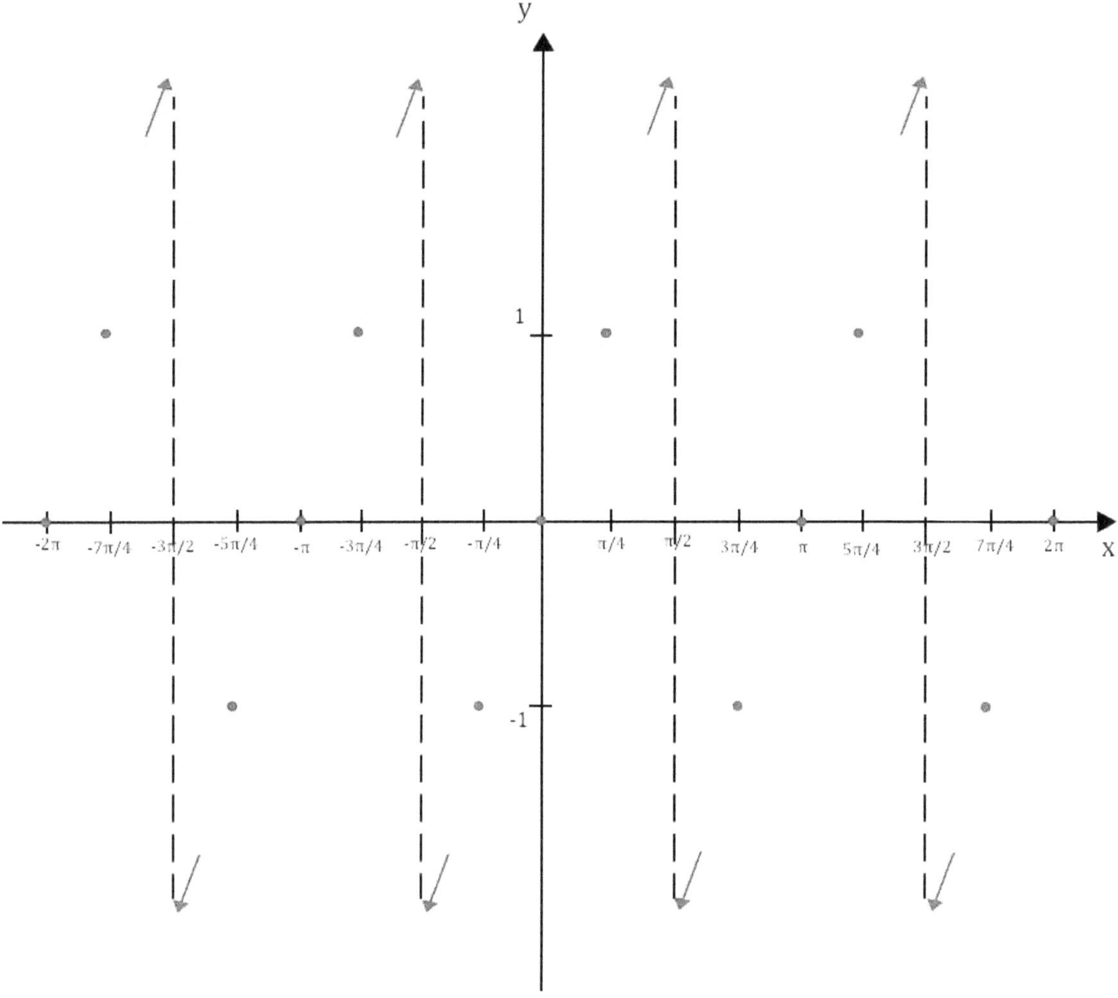

Figure 5.49

Now by connecting the dots in a smooth way you would get the following shape for tan(x). What is the smooth way? Good question. For now, try to be gentle, no sharp turns! By doing so, you would get a repeating graph that represents tan(x) at different angles. Please note that your graph may look a bit different as we have tried to show more points on the y-axis to make it smoother.

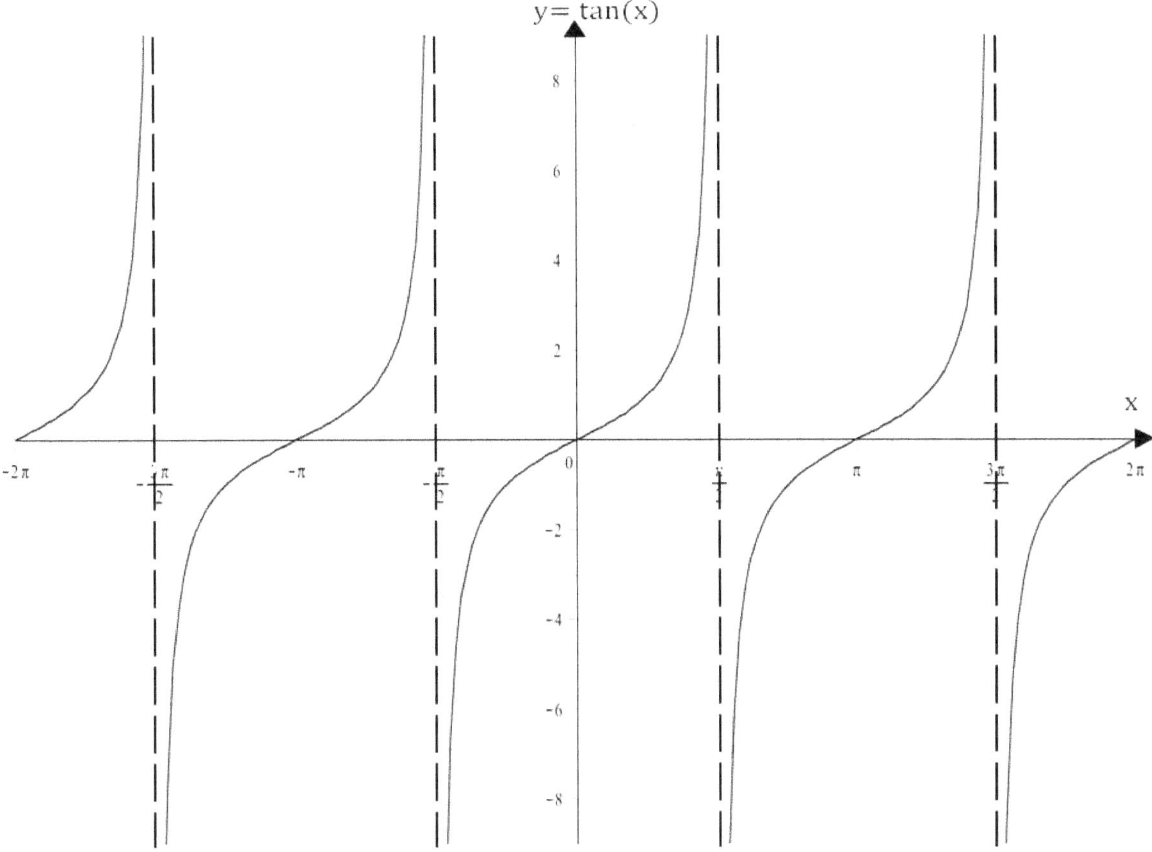

Figure 5.50

We can use the same analogy to graph cot (x). To do so, all we require is to find the associated cotangent values for different angles. Since cot(x) is defined as follows:

$$\cot(x) = \frac{\cos(x)}{\sin(x)}$$

We can see the values that make sin(x) = 0; such as $0, \pi, 2\pi, -\pi,$ and -2π, would make denominator zero. Therefore, they are possible vertical asymptotes, and they can be used as a guide to draw cotangent function in the following presented manner:

Why possible vertical asymptotes?

In our case, they are vertical asymptotes, but in general, it is not always true. That is, we have to make sure the values that make denominator zero are not making the numerator zero as well. If they make the numerator zero as well, the points are most likely discontinuity points rather than vertical asymptotes.

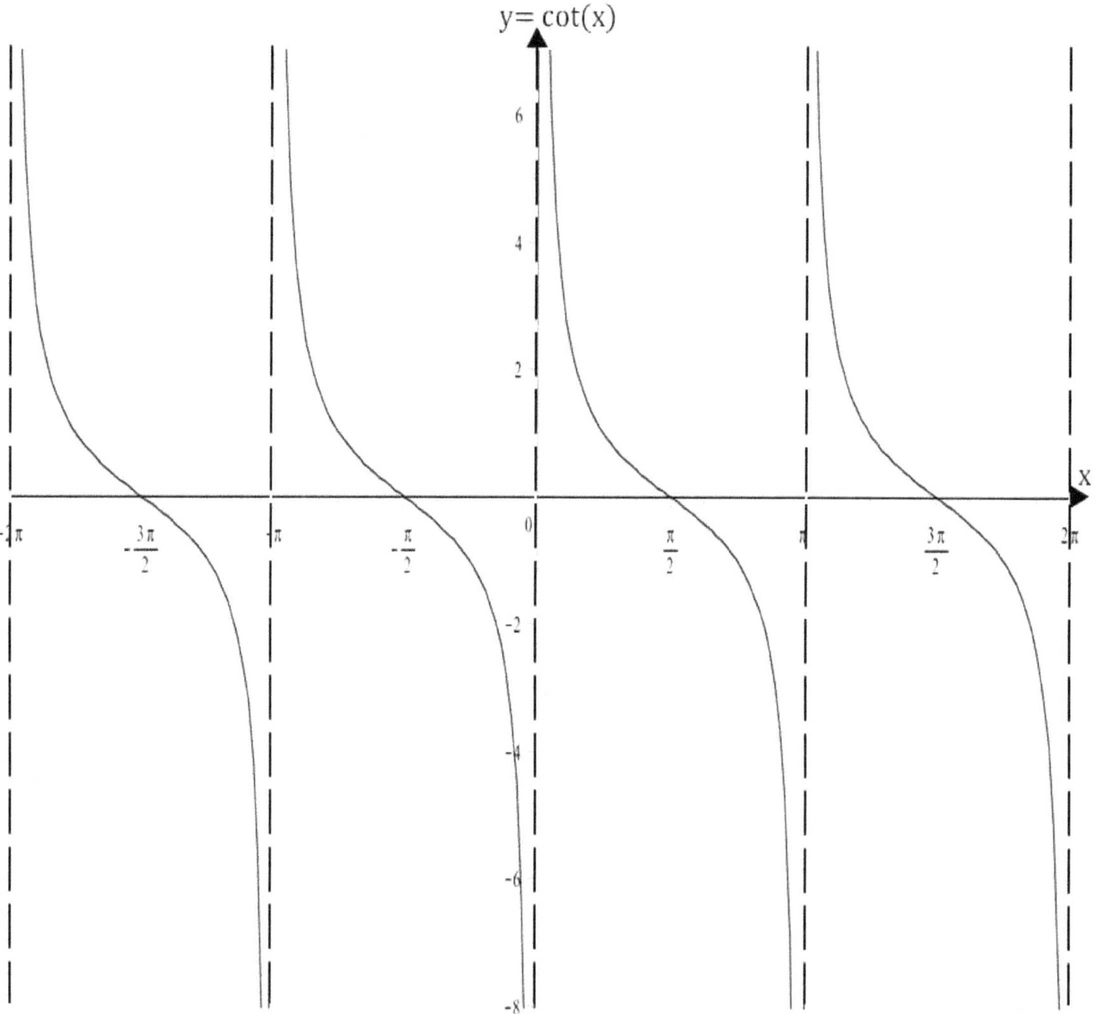

Figure 5.51.

5.8. Period of tan(x) and cot(x)

As you might have noticed, we can obtain the period of the tan(x) and cot(x) from the graphs. The graphs that we have displayed demonstrate the periods for the tan(x) and cot(x) are equal to π. That is, for every π, the tan(x) and cot(x) function would repeat themselves. Please refer to the following illustrations for further clarification.

y = tan(x)

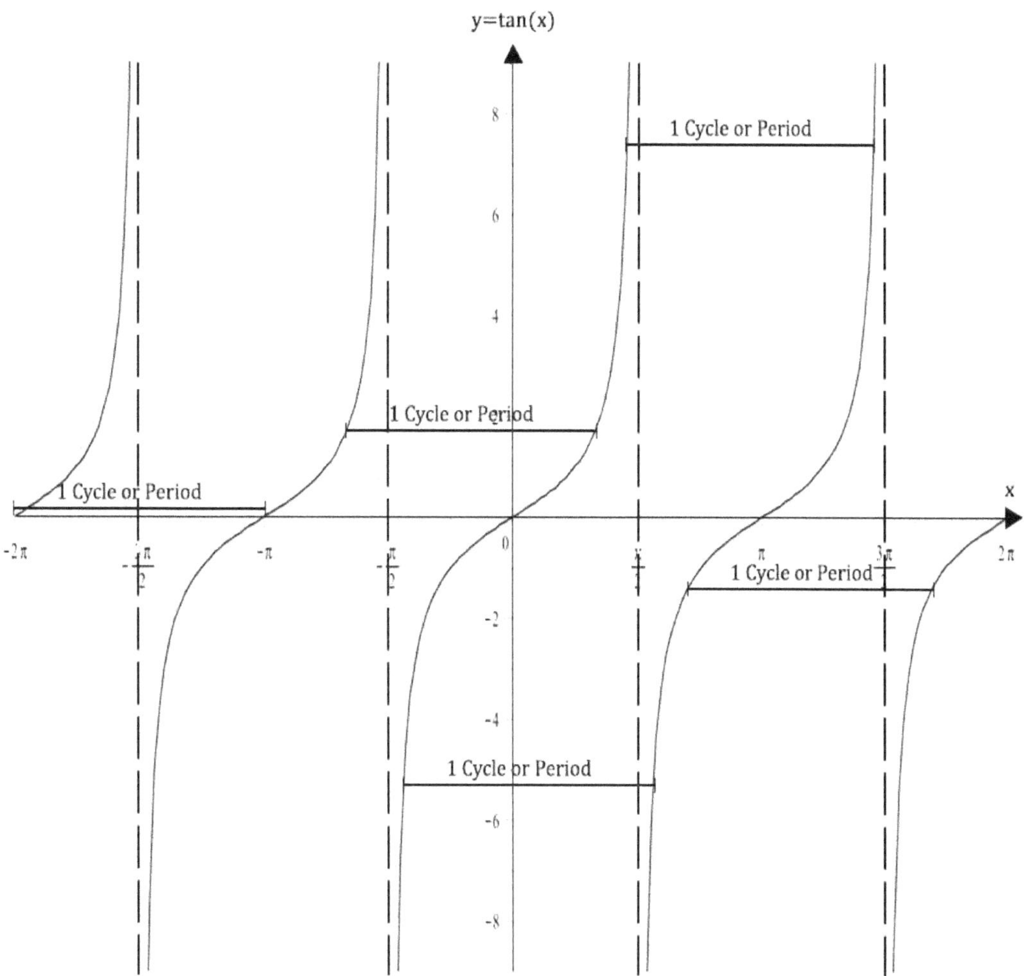

Figure 5.52.

y = cot(x)

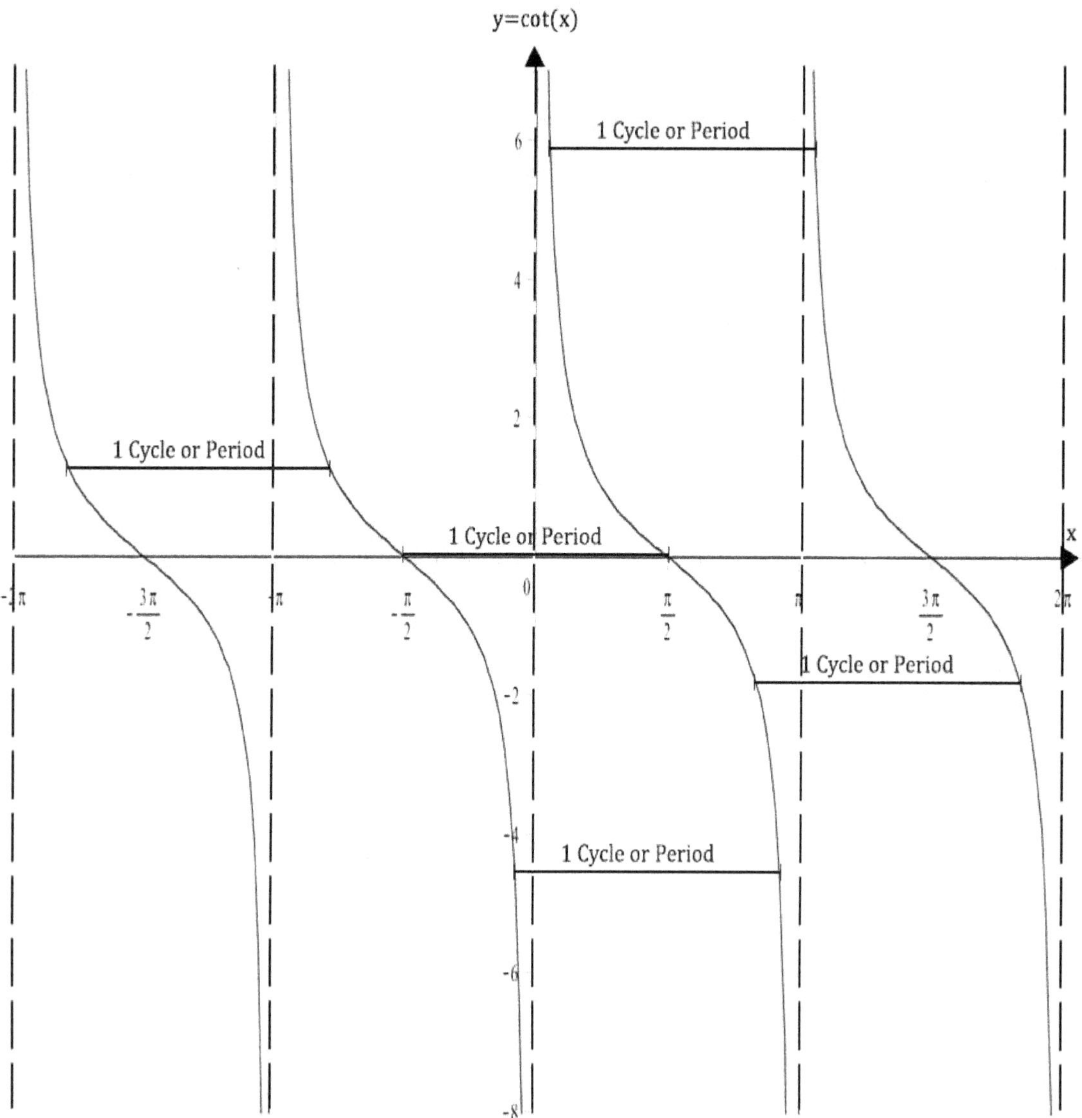

Figure 5.53.

5.9. Transformation of tan(x) function

There is a general formula that we can use to define the tan(x) function, which is presented as follows:

$$y = A\tan\big(b(x-c)\big) + d$$

In the presented formula A, b, c and d can be defined as follows:

"A" would be amplitude, and it is the distance from the median line to the peak (highest point) or distance from the median line to the trough. In simpler words, it is the amount of vertical compression or stretch. That is, if A is 3, then you would stretch it three times vertically.

"c" is the phase shift, horizontal shift or horizontal translation. It is how much you would move back and forth. Please note the negative sign behind c. That is, if c is positive, you will get a negative number in the formula, and, if c were negative, you would get a positive number. If c is positive, you have to move the function to the right, and, if negative, you have to shift to the left.

"d" is the vertical shift or vertical translation. It states how much you should move the entire graph up or down. Another analogy would be how much you should move the median line up or down. For example, if d is 4, the entire graph would be moved up by 4 units.

"b" can tell us about the period, but it is NOT the period. To find the period, we use b in the following formula:

$$\frac{\pi}{b} = period\ or\ \frac{\pi}{period} = b$$

One complete cycle. That is, you start from a certain point (y-value) and a certain direction, and you get back to the same point while keeping the same direction.

Now let's try a few questions about transformation of sin(x) function, given that we know how to graph y = tan(x). Please refer to the following diagram for clarification.

I believe it is better if we graph the functions in the suggested stepwise manner, however:

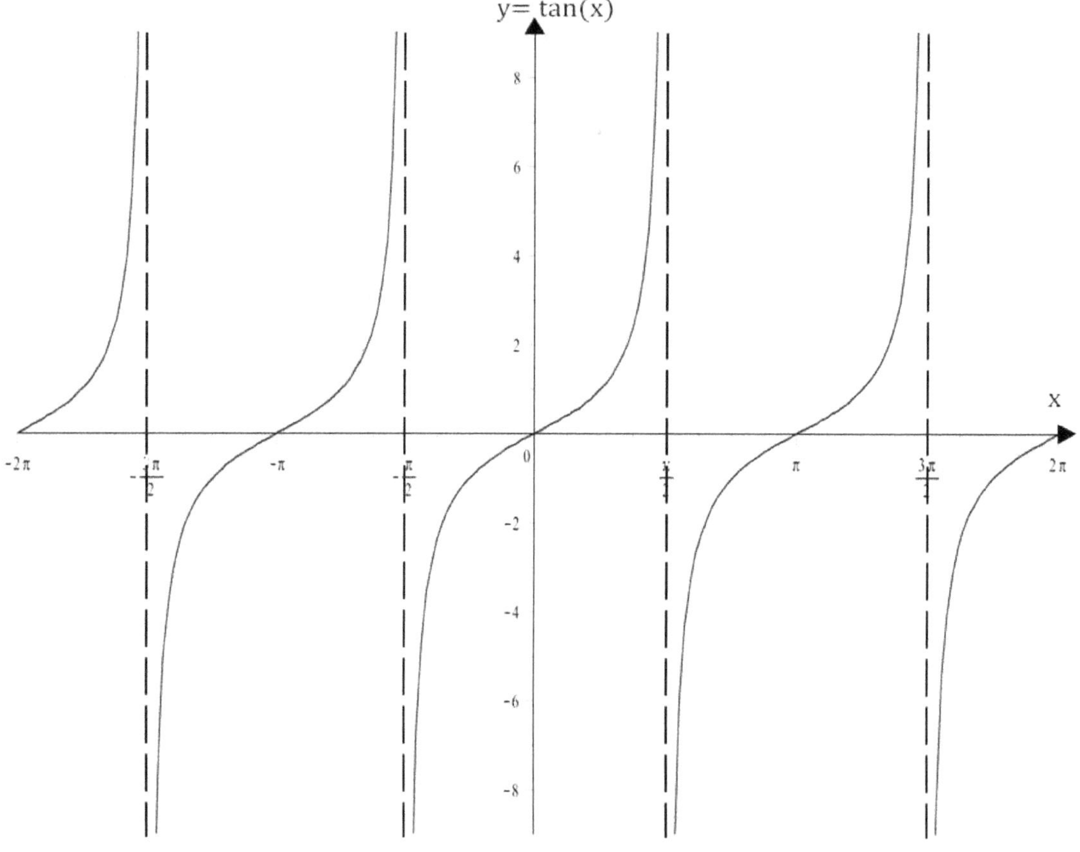

Figure 5.54.

You may notice there are other possibilities to get to the right answer. You can find the list of steps that I found easy to follow.

1. First, try to put the function in the format that we have presented, which is: $y = A\tan\big(b(x-c)\big)+d$.

2. Start with graphing the y = tan(x) function.
3. Always start inside out, that is, first deal with "c", the phase shift (move the tan(x) function to the left or right).
4. Then you take care of "b" to find the period. The period would let us know the length of our function in the x-direction for one complete cycle.
5. The next would be considering amplitude. For the case of tan(x) or cot(x), it means how much you would stretch or compress the function vertically.
6. The vertical translation is the last step that I recommend. Please note you can do it in many other fashions, and you may still get the correct answer. Use your judgment to find your way to draw.

We now use the steps to draw the graph of the function below:

$$y = 2(\tan(3x - 2\pi)) - 1$$

1. The first step is to put the function in the standard form. The only thing that we need to do is factoring 3, in which case we get:

$$y = 2\tan\left(3\left(x - \frac{2\pi}{3}\right)\right) - 1$$

2. We then graph $y = \tan(x)$.

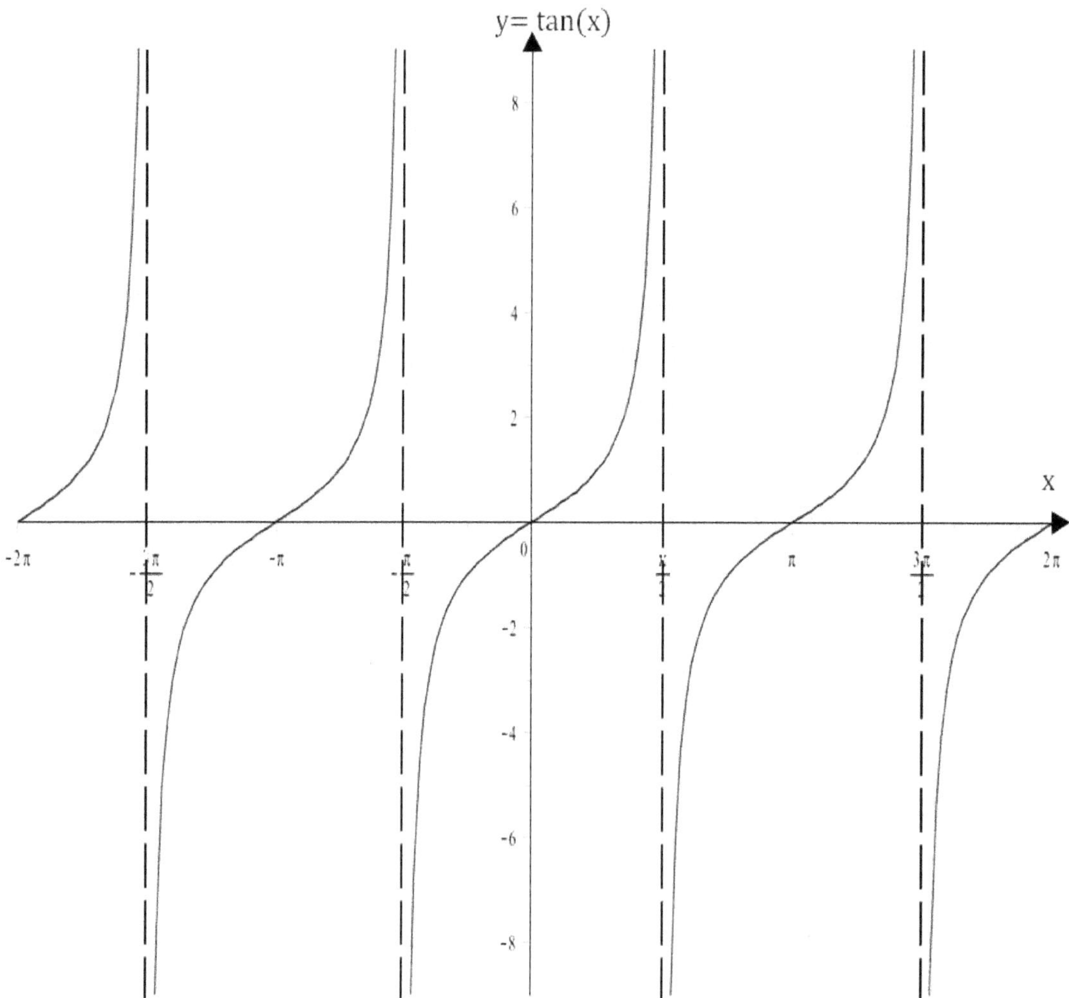

Figure 5.55.

3. To graph $y = \tan\left(\left(x - \frac{2\pi}{3}\right)\right)$, we need to move the tan(x) to the right by $\frac{2\pi}{3}$ unit. Notice c is $\frac{2\pi}{3}$, which we interpret as moving the function $\frac{2\pi}{3}$ to the right.

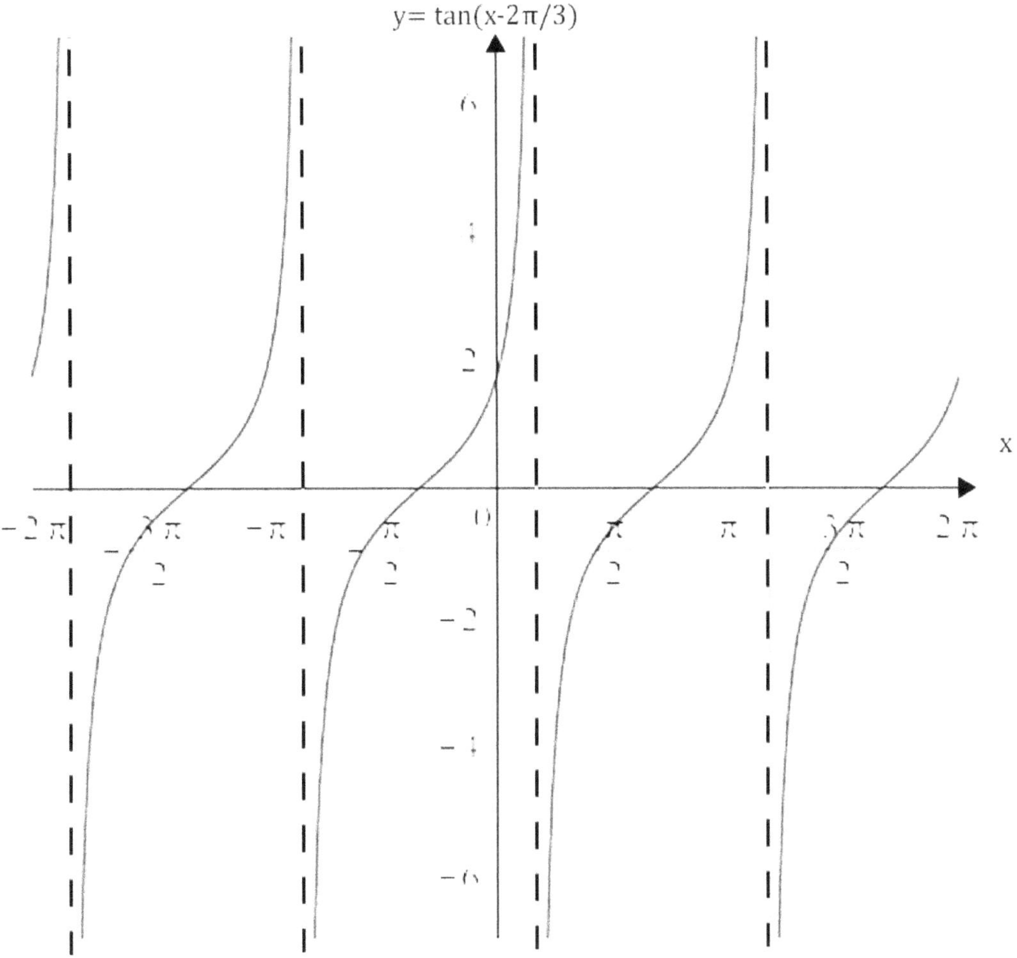

Figure 5.56.

4. To graph $y = \tan\left(3\left(x - \frac{2\pi}{3}\right)\right)$, we have two paths that we can take, first by squeezing the function by three since "b" is 3. The other way to approach is to find the period by saying b=3. We use the following formula to obtain the period:

$$\frac{\pi}{b} = period => \frac{\pi}{3}$$

Then we would get the following graph:

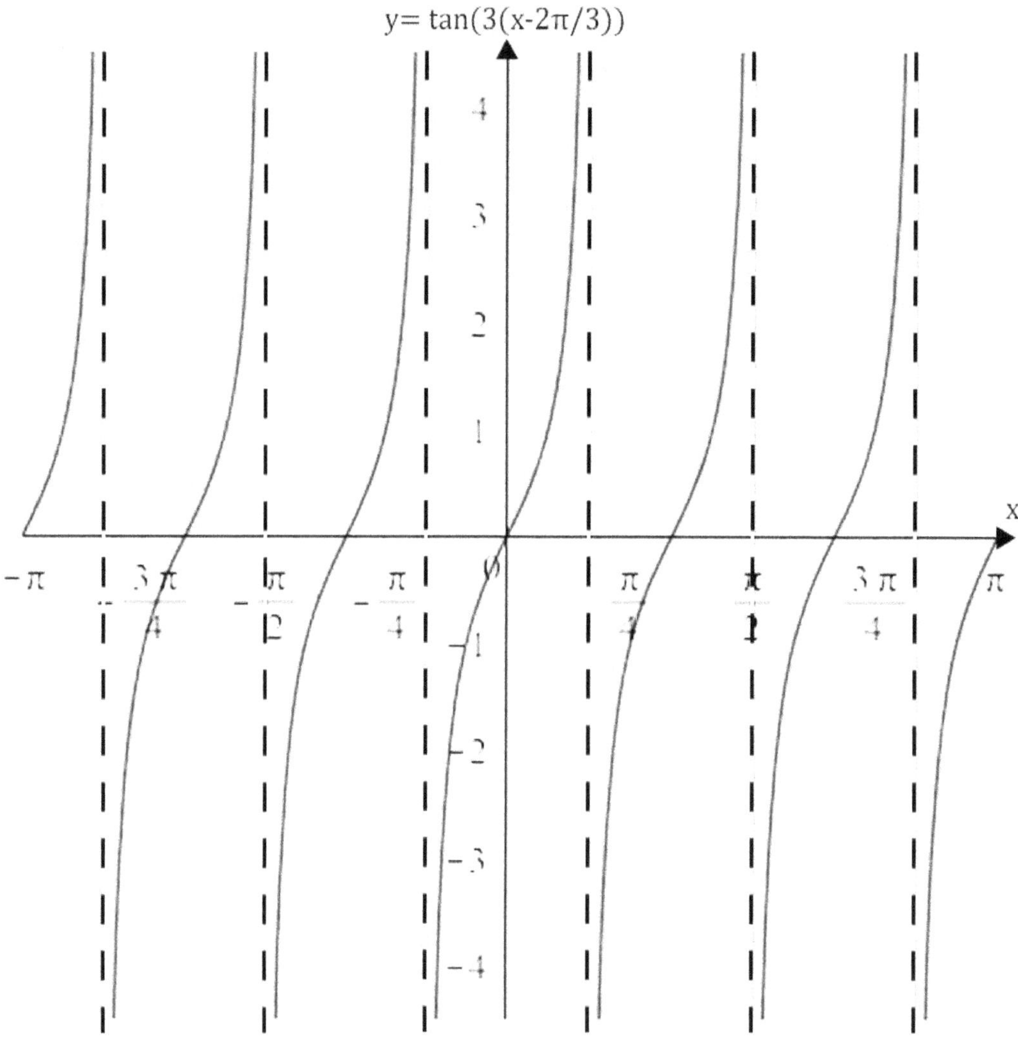

Figure 5.57.

5. The next step would be to graph $y = 2\tan\left(3\left(x - \frac{2\pi}{3}\right)\right)$. With the knowledge that the amplitude "A" is equal to 2, we know we have to stretch the function by two from both the top and bottom.

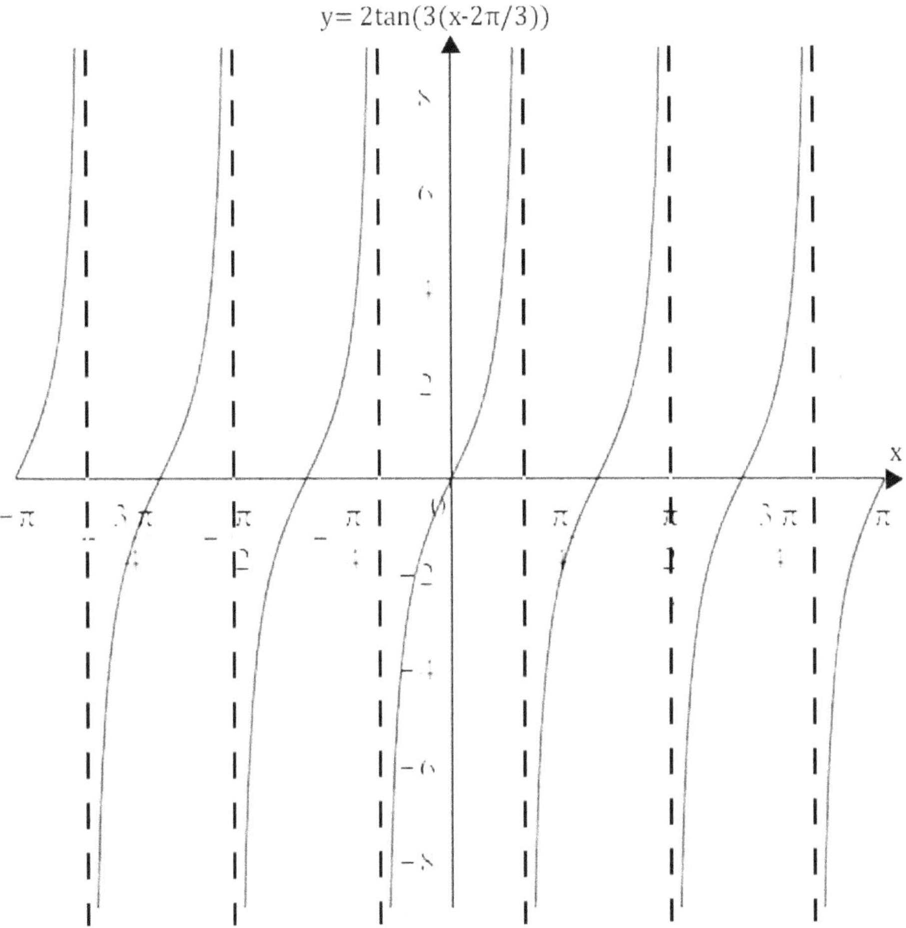

Figure 5.58.

6. The final step is to graph $y = 2\tan\left(3\left(x - \frac{2\pi}{3}\right)\right) - 1$. Knowing "d" is -1, it means we need to move the function one unit down. That is, the new median line would be at y = -1 instead of y = 0 (the x-axis).

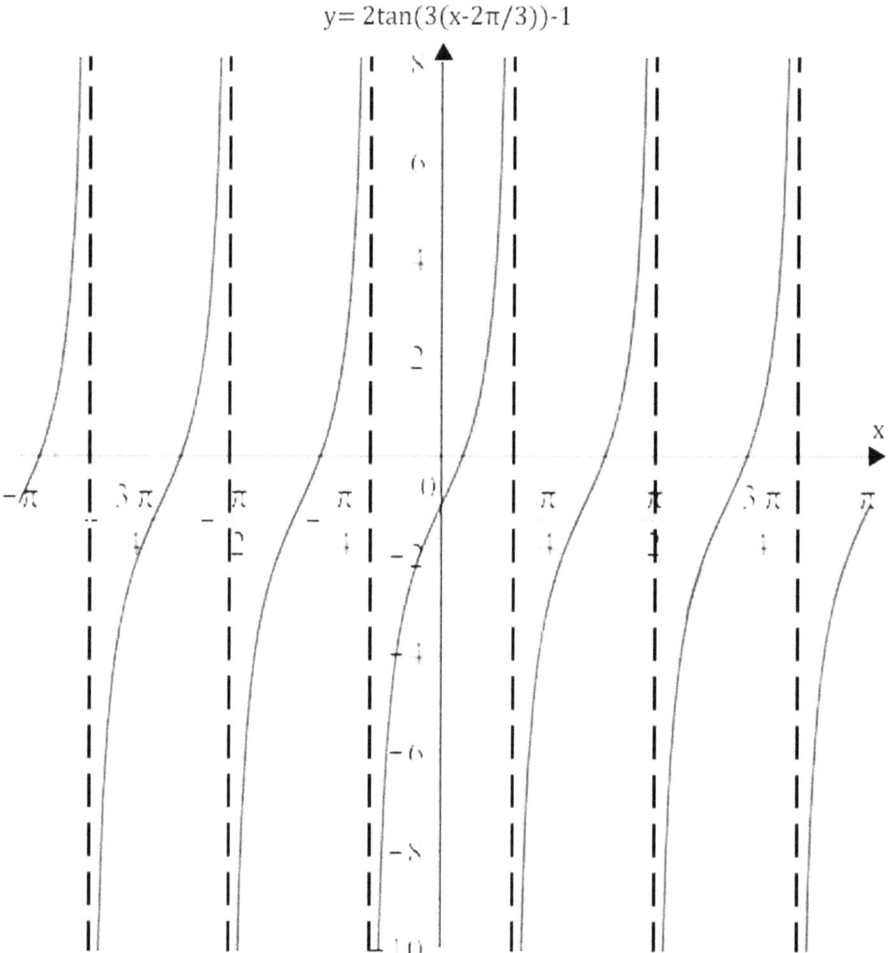

Figure 5.59.

5.10. Transformation of cot(x) Function

There is a general formula that we can use to define the cot(x) function, which is presented as follows:

$$y = A\cot\big(b(x - c)\big) + d$$

In the presented formula A, b, c and d can be defined as follows.

"A" would be amplitude, and it is the distance from the median line to the peak (highest point) or distance from the median line to the trough. In simpler words, it is the amount of vertical compression or stretch. That is, if A is 3, then you would stretch it three times vertically.

"c" is the phase shift, horizontal shift or horizontal translation. It is how much you would move back and forth. Please note the negative sign behind c. That is, if c is positive, you will get a negative number in the formula, and if c were negative, you would get a positive number. If c is positive, you have to move the function to the right, and, if negative, you have to shift to the left.

"d" is the vertical shift or vertical translation. It states how much you should move the entire graph up or down. Another analogy would be how much you should move the median line up or down. For example, if d is 4, the entire graph would be moved up by 4 units.

"b" can tell us about the period, but it is NOT the period. To find the period, we use b in the following formula:

$$\frac{\pi}{b} = period\ or\ \frac{\pi}{period} = b$$

One complete cycle. That is you start from a certain point (y-value) and a certain direction, and you get back to the same point while keeping the same direction.

Now let's try a few questions about transformation of cot(x) function given that we know how to graph y = cot(x). Please refer to the following diagram for clarification:

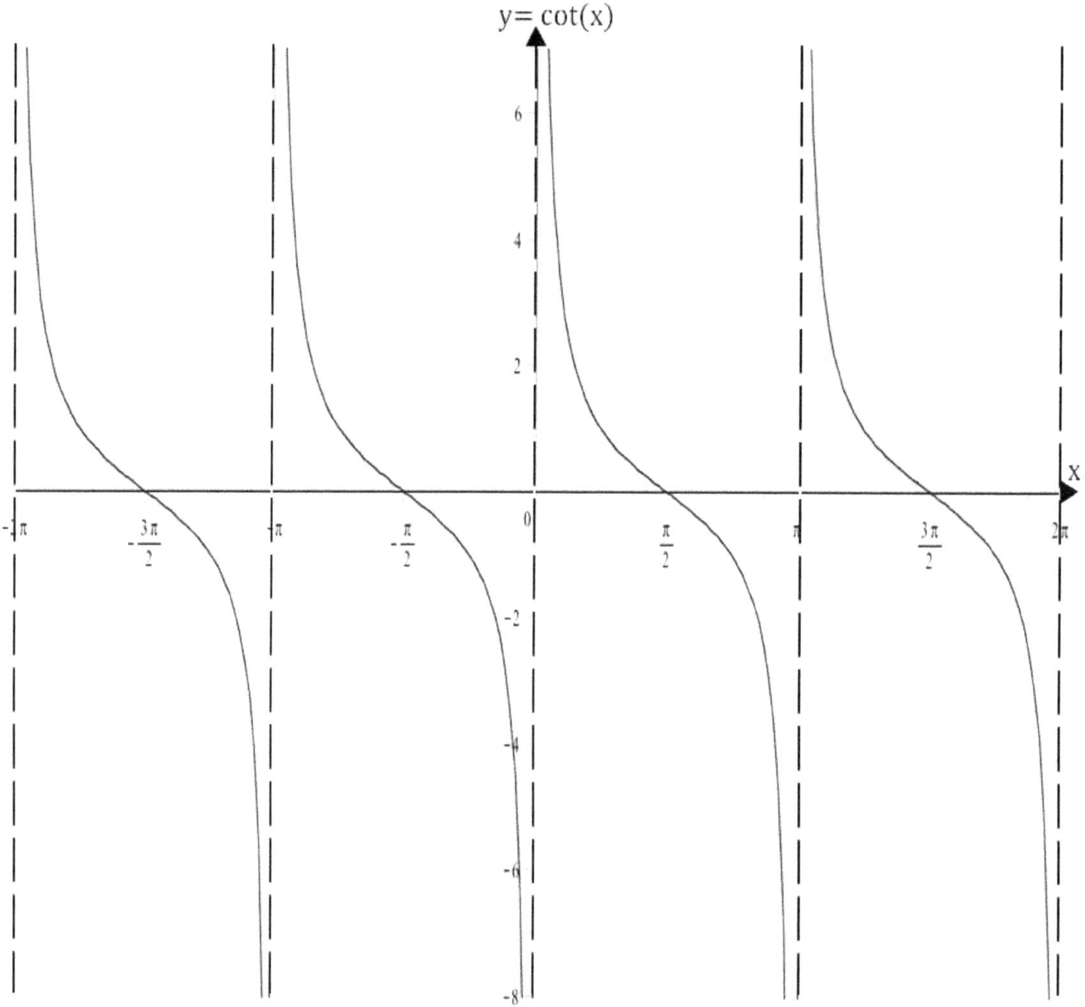

Figure 5.57.

I believe it is better if we graph the functions in the suggested stepwise manner; however, you may notice there are other possibilities to get to the right answer. You can find the list of steps that I found easy to follow.

1. First, try to put the function in the format that we have presented, which is: $y = A\cot\left(b(x-c)\right)+d$.
2. Start with graphing the y = cot(x) function.
3. Always start inside out, that is, first deal with “c”, the phase shift (move the cot(x) function to the left or right).
4. Then you take care of "b" to find the period. The period would let us know the length of our function in the x-direction for one complete cycle.
5. The next would be considering amplitude. For the case of tan(x) or cot(x), it means how much you would stretch or compress the function vertically.
6. The vertical translation is the last step that I recommend. Please note you can do it in many other fashions, and you may still get the correct answer. Use your judgment to find your way to draw.

We now use the steps to draw the graph of the function below.

$$y = -\frac{1}{3}\cot(2x-\pi)+2$$

1. The first step is to put the function in the standard form. The only thing that we need to do is factoring 2, in which case we get:

$$y = -\frac{1}{3}\cot\left(2\left(x-\frac{\pi}{2}\right)\right)+2$$

2. We then graph $y = \cot(x)$

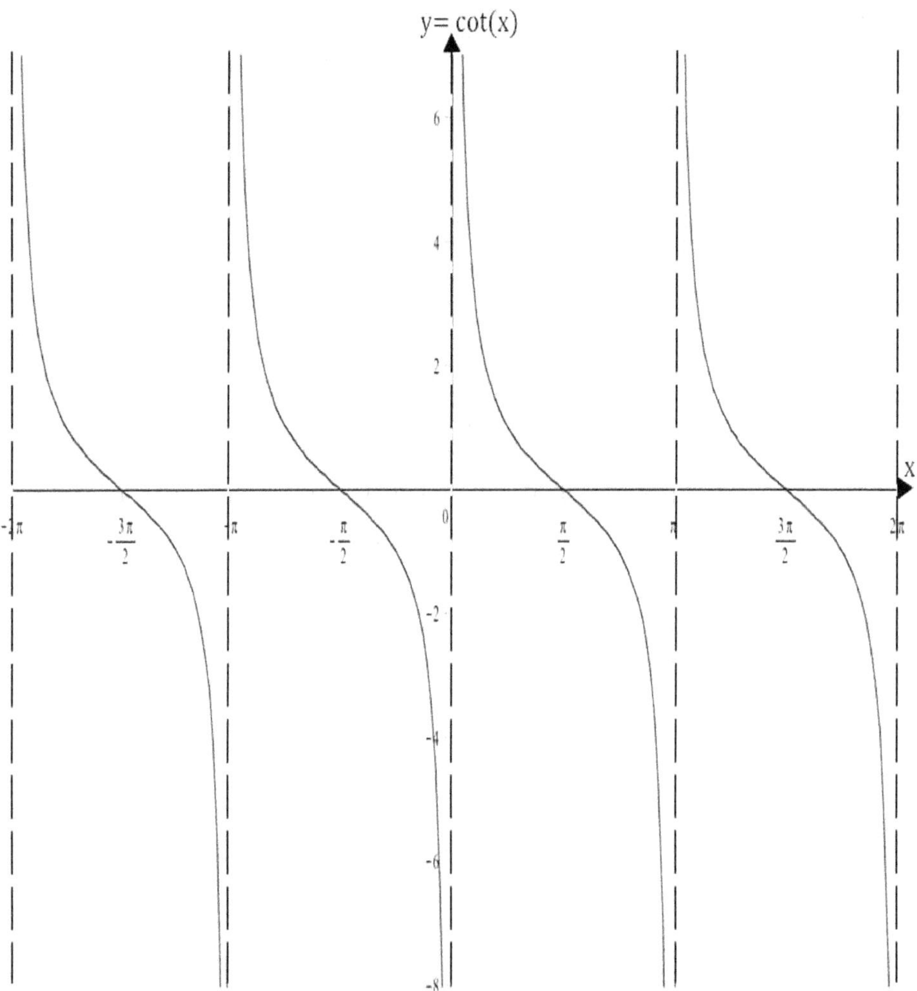

Figure 5.58.

3. To graph $y = \cot\left(\left(x - \frac{\pi}{2}\right)\right)$, we need to move the tan(x) to the right by $\frac{\pi}{2}$ unit. Notice c is $\frac{\pi}{2}$, which we interpret as moving the function $\frac{\pi}{2}$ to the right.

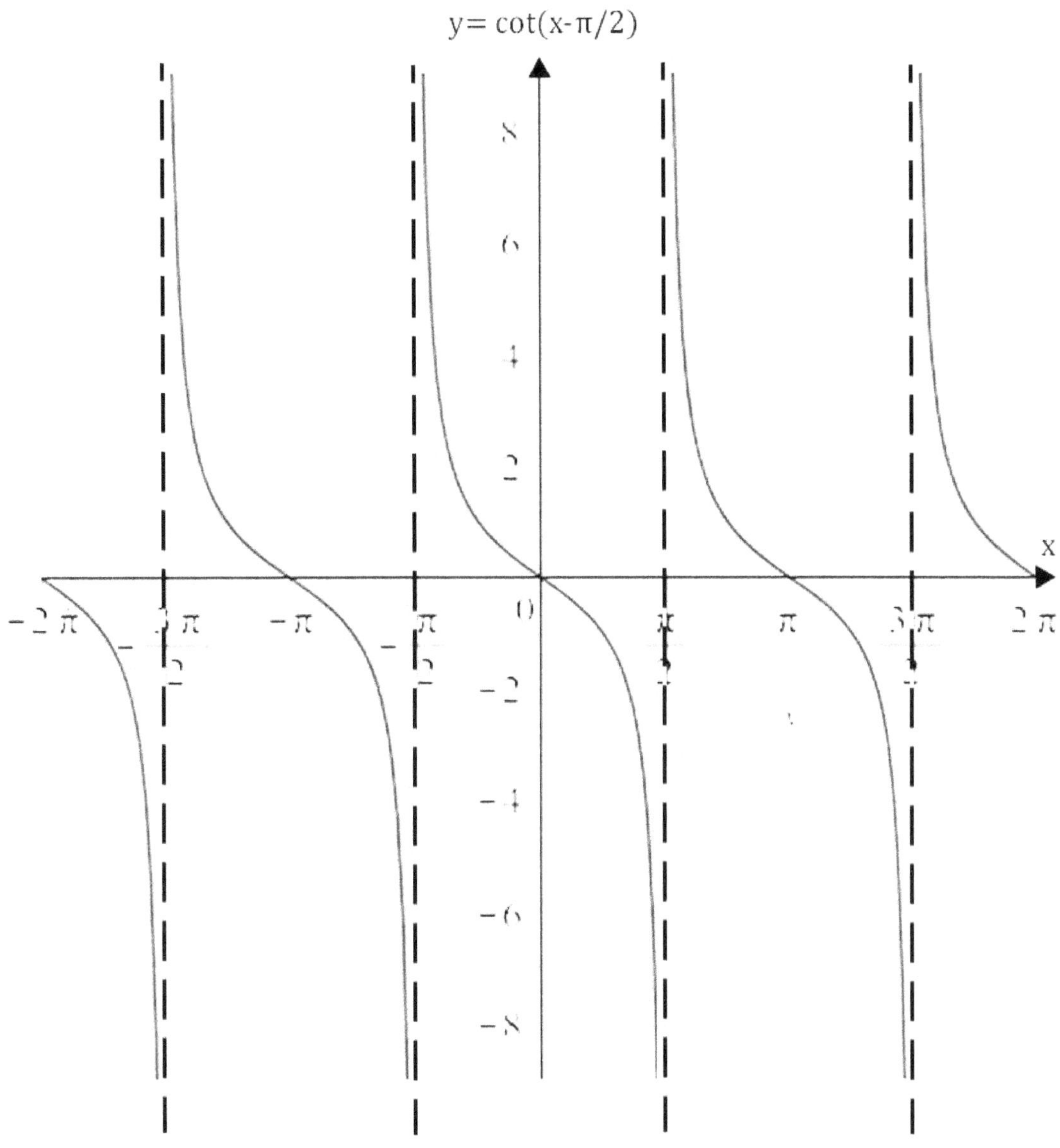

Figure 5.59.

4. To graph $y = \cot\left(2\left(x - \frac{\pi}{2}\right)\right)$, we have two paths that we can take, first by squeezing the function by two since "b" is 2. The other way to approach it is to find the period by saying b=2. We use the following formula to obtain the period:

$$\frac{\pi}{b} = period => \frac{\pi}{2}$$

Then we would get the following graph:

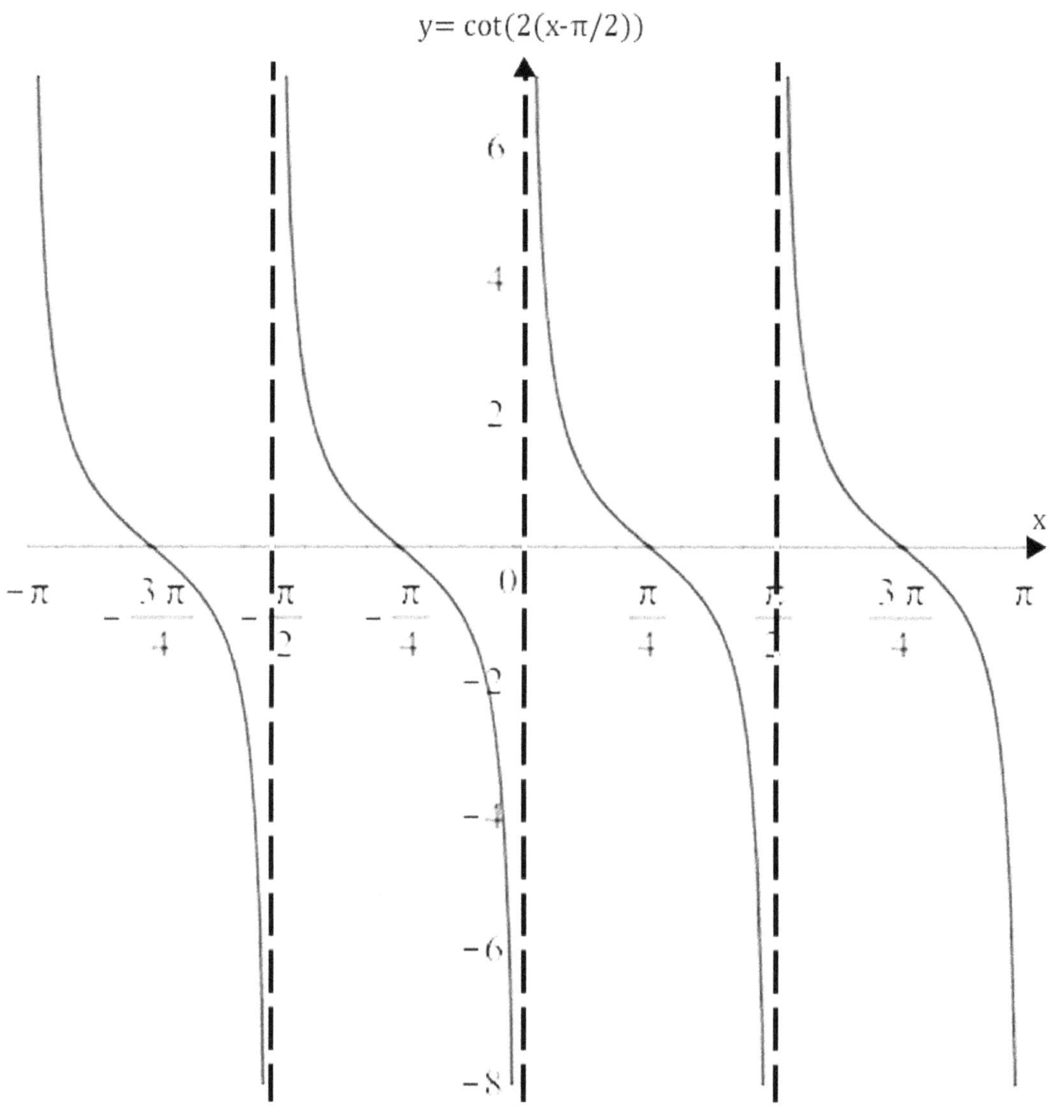

Figure 5.59.

5. The next step would be to graph $y = -\frac{1}{3}\cot\left(2\left(x - \frac{\pi}{2}\right)\right)$. With the knowledge that the amplitude "A" is equal to $-\frac{1}{3}$, we know we have to squeeze the function by three vertically. After that, the negative sign of the amplitude is an indication that we need to flip the function about the x-axis.

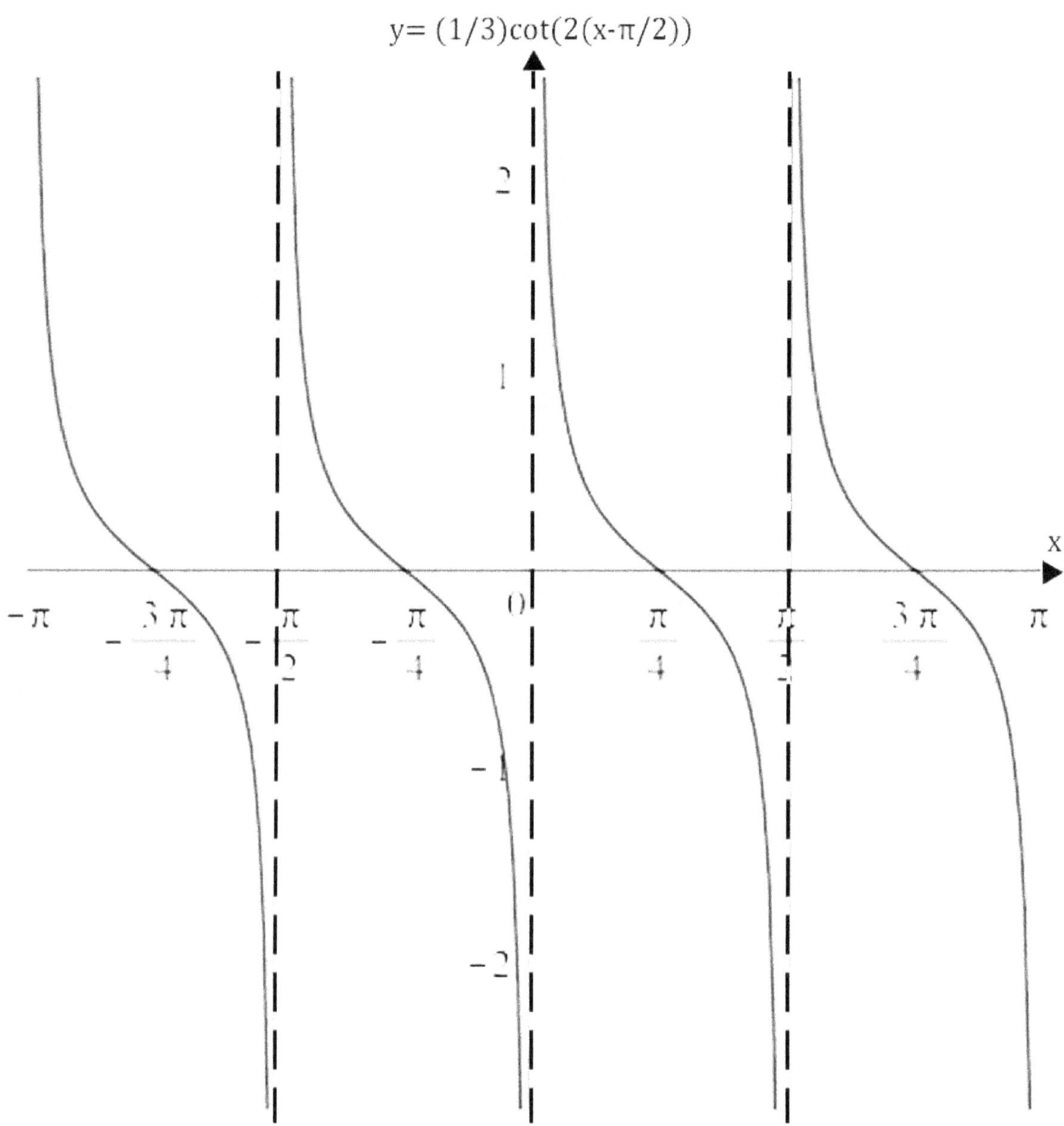
y= (1/3)cot(2(x-π/2))
x
2
1
0
-1
-2
-π
$-\frac{3\pi}{4}$
$-\frac{\pi}{2}$
$-\frac{\pi}{4}$
$\frac{\pi}{4}$
$\frac{\pi}{2}$
$\frac{3\pi}{4}$
π

Figure 5.60.

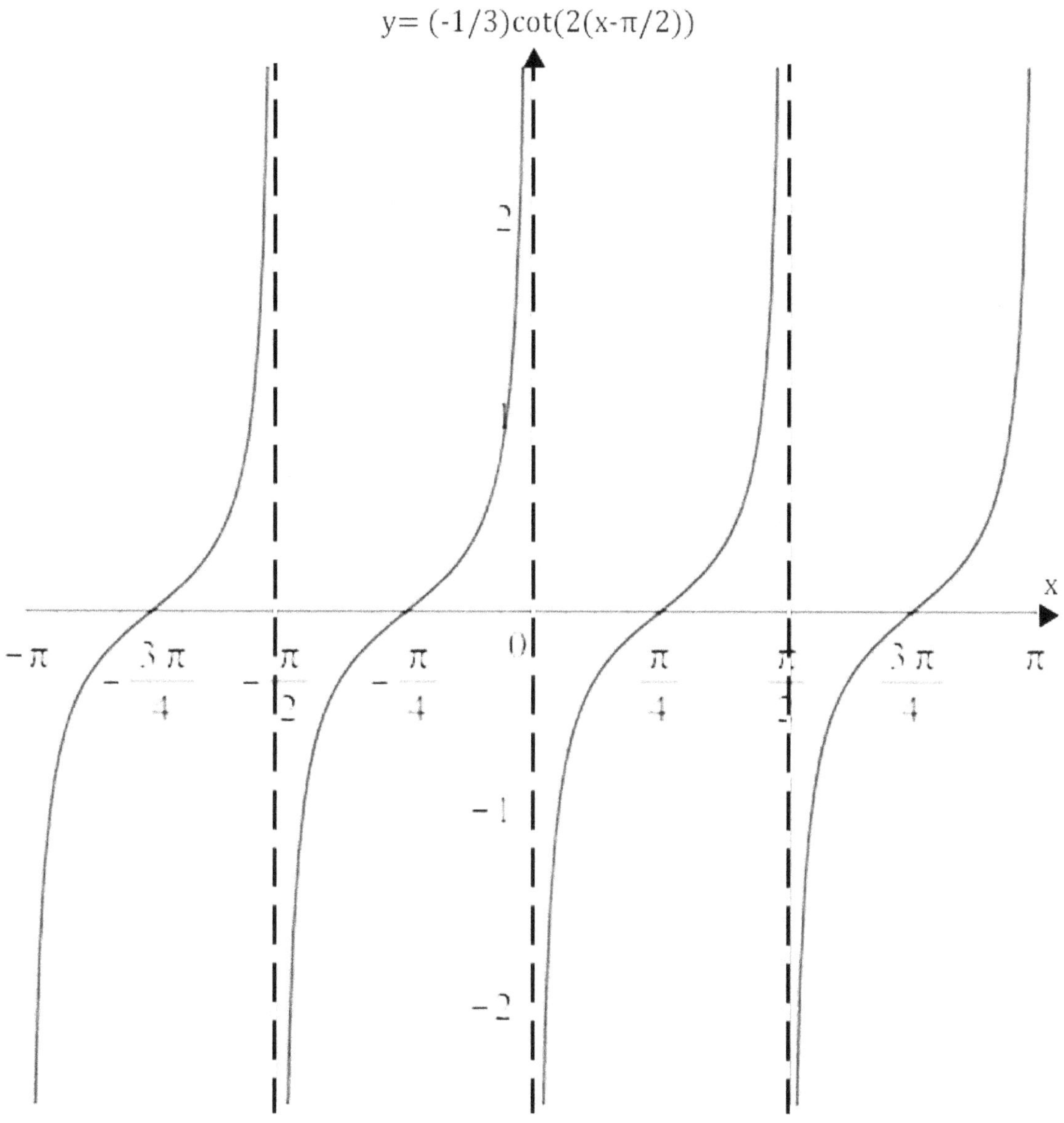

Figure 5.61.

6. The final step is to graph $y = -\frac{1}{3}\cot\left(2\left(x - \frac{\pi}{2}\right)\right) + 2$. Knowing "d" is 2, it means we need to move the function two units up. That is, the new median line would be at y = +2 instead of y = 0 (the x-axis).

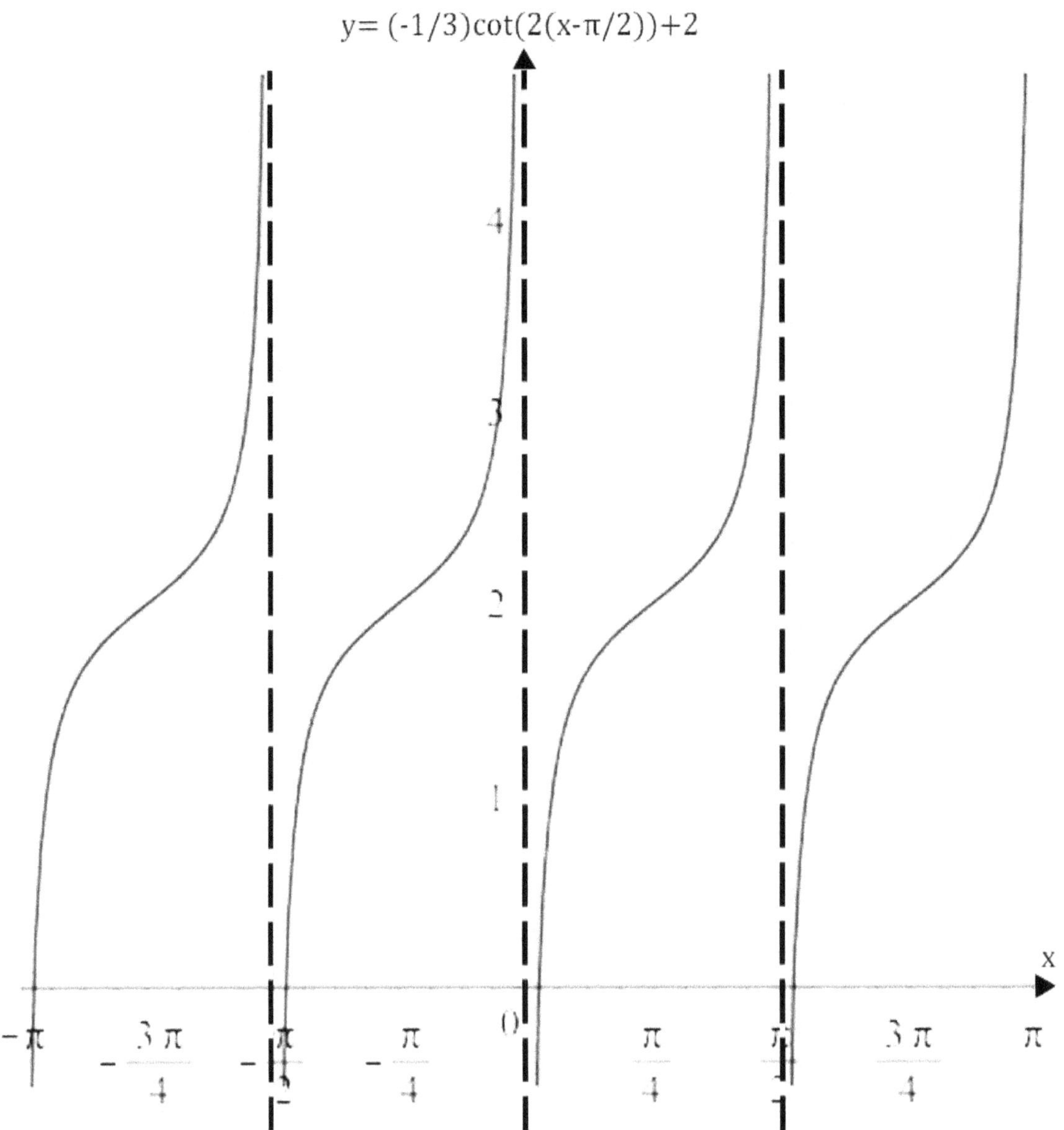

Figure 5.62.

6. Inverse Trigonometric Functions

I love watermelon, and I would assume a machine that produces watermelon juice is great to have. Quite honestly, I am not sure if the watermelon juice is going to be as good as I think. Having said that, I prefer a machine that helps me to turn the watermelon juice back into the watermelon again. That machine is worth a lot, and I am ready to call it Awesome 33 Machine (the reason I called it Awesome 33 Machine is because the seedless watermelon has 33 chromosomes inside their white coats). Now the first machine, which turns watermelon to the juice, can be defined as "function." On the other hand, the Awesome 33 Machine that turns the watermelon juice into watermelon would be called the "inverse." You may argue that the Awesome 33 Machine is practically impossible, even though the first device exists. In that case, you are right, not every function has an inverse.

There are at least three possibilities that can occur. First is that the inverse does not exist. The second scenario is that the inverse of the function exists, but it is not a function itself. The third possibility is that the inverse exists, and it is a function itself. If you are asking what a function is, the simple answer is, I expect to get only one type of juice from any fruit that I put in the machine (function). Hence, if I put orange, I am not expecting to get either orange juice or apple juice. If such a thing happens, that machine is no longer a function, and we call it a relation.

In mathematics, it is a common practice to show a function with $f(x)$, which we can read as "f" is a function of x. In more simple terms, the fruits that the machine accepts (variables) are represented by x. In this case, the inverse of a function is shown using $f^{-1}(x)$. Please note that $f^{-1}(x)$ does not mean $\frac{1}{f(x)}$.

6.1. Inverse of sin(x)

So far we have explained what we mean by saying $\sin\left(\frac{\pi}{6}\right) = x$. We expect the value of x to be a ratio between -1 and 1, inclusively. In this particular case, the answer is 0.5. However, what should we do if we are interested to know the sine of what angles are equal to the ratio of 0.5. The answer is we need to use the inverse function. In the case of sine, if we show $f(x) = \sin(x)$, then we demonstrate the inverse of sine as $f^{-1}(x) = \sin^{-1}(x) = \arcsin(x)$. Please note there are two standard notations for showing the inverse of trigonometric function $\sin^{-1}(x)$ and $arcsin(x)$.

Before going into detail, let's clarify an issue with the notation. There is confusion associated with the first notation. You may have seen, in the past, that $\mathrm{x}^{-1} = \frac{1}{x}$; however, we should not confuse this argument with $\sin^{-1}(x)$. In other words, we should keep in mind the following statement is true:

$$\mathbf{\sin^{-1}(\boldsymbol{x}) \neq \frac{1}{\sin(\boldsymbol{x})}}$$

In case we wanted to show $\frac{\mathbf{1}}{\mathbf{sin}(x)}$, we can use the following notation:

$$\mathbf{(\sin(x))^{-1} = \frac{1}{\sin(\boldsymbol{x})}}$$

Now let's see what $\sin^{-1}(x)$ or $\arcsin(x)$ means. How about using an example? $\mathrm{Sin}^{-1}(1)$ means the sine of what angle is 1. Therefore, the result of the inverse trigonometric function should be an angle. Please note there are many possible correct answers for the angles. For example, $\sin^{-1}(.5)$ can be any of the following angles:

$$\frac{\pi}{6}, \frac{5\pi}{6}, \frac{13\pi}{6}, \frac{17\pi}{6}, \frac{25\pi}{6}, \frac{29\pi}{6}, \ldots$$

Alternatively, in mathematical terms, we can state: $2k\pi + \frac{\pi}{6}$ union with $2k\pi + \frac{5\pi}{6}$ where $k = 0, \pm 1, \pm 2, \pm 3, \ldots$

In fact, the truth is this; every single trigonometric inverse would have infinitely many possible answers. Now there should be a few questions arising.

First of all, if the inverse is a function, shouldn't we get only one answer? The short answer is yes. The next question can be, which one should we write out of infinitely many? The short answer is there is no short answer. There are three possible strategies to respond these questions.

The first solution is, list all the answers as we did. For example, for $\sin^{-1}(.5) = \frac{\pi}{6}, \frac{5\pi}{6}, \frac{13\pi}{6}, \frac{17\pi}{6}, \frac{25\pi}{6}, \frac{29\pi}{6}, \ldots$. It is possible to use the symbolic notation as follows: $\sin^{-1}(.5) = \left(2k\pi + \frac{\pi}{6}\right) \cup \left(2k\pi + \frac{5\pi}{6}\right)$ where $k = 0, \pm1, \pm2, \pm3, \ldots$ (The ∪ mean "or." Basically, we want both sets of values.)

If you use the former approach, you must keep in mind that your result represents an inverse relation rather than a function.

The second approach is, by convention, if no range is defined for the angle, the result of sine inverse function would always lie between $-\frac{\pi}{2}$ and $\frac{\pi}{2}$. Also, it can be written as follows:

$$\{\sin^{-1}(x) = y \mid x, y \in \mathbb{R}, -\frac{\pi}{2} \leq y \leq \frac{\pi}{2}, -1 \leq x \leq 1\}$$

The question can be why between $-\frac{\pi}{2} \leq y \leq \frac{\pi}{2}$?

Lastly, it is possible that we choose any arbitrary interval that would cover all possible values for sin(x), i.e., from -1 to +1. One possible solution is $-\frac{\pi}{2} \leq y \leq \frac{\pi}{2}$, the other one could be $\frac{\pi}{2} \leq y \leq \frac{3\pi}{2}$. The former interval is more widely accepted; that is why we use $-\frac{\pi}{2} \leq y \leq \frac{\pi}{2}$ as the range. The following graph of arcsin(x) should illustrate why domain and range of arcsin(x) is as they were defined.

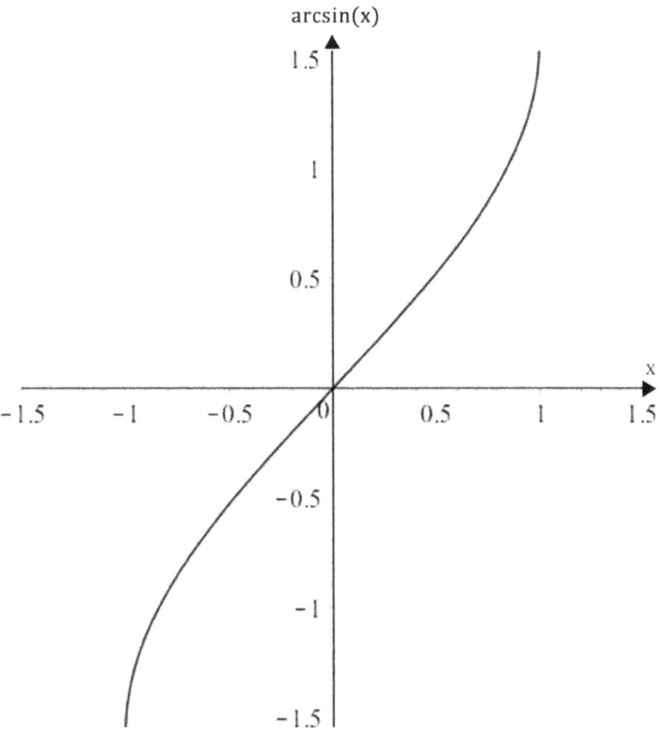

Figure 7.1.

F.A.Q.

---

1. Is there any other possible way to choose the range?

Yes, there are infinitely many possibilities.

2. Why not from $-\pi \leq y \leq \pi$?

In that case, we would have had duplicated values, so our inverse would not have been a function. As an example:

$$\sin^{-1}\left(\frac{1}{2}\right) = y => y = \frac{\pi}{6} \; or \; \frac{5\pi}{6}$$

In that case, one value of x, namely $\frac{1}{2}$, led to two values of y ($\frac{\pi}{6}$ or $\frac{5\pi}{6}$) since we chose y to be between$-\pi \leq y \leq \pi$. However, if we choose the range to be $-\frac{\pi}{2} \leq y \leq \frac{\pi}{2}$, the aforementioned problem will not occur, and the only answer we obtain would be $\frac{\pi}{6}$. Also, such range for arcsin(x) would not cover the negative values for sin(x).

3. Setting the range in a way that we only obtain one answer does not guarantee that we have the correct answer. How do we know we have not eliminated the right answer due to range restriction?

In that case, we should use the information in the question to figure out if the obtained solution is accurate. Remember, we can always use the following trigonometric identity:

$$\sin(x) = \sin(\pi - x)$$

Using the aforementioned formula, we can always find the other solutions. For instance, in the prior question, we are supposed to locate the value of $\sin^{-1}\left(\frac{1}{2}\right) = y$ in a way the solution resides in the second quadrant. In that case, the following is easy to obtain:

$$\sin^{-1}\left(\frac{1}{2}\right) = y => y = \frac{\pi}{6}$$

The obtained solution is in the first quadrant. To find the answer in the second quadrant, we use the identity $\sin(x) = \sin(\pi - x)$. Therefore, the angle in the second quadrant would be:

$$\sin\left(\frac{\pi}{6}\right) = \sin\left(\pi - \frac{\pi}{6}\right) => \sin\left(\frac{\pi}{6}\right) = \sin\left(\frac{6\pi}{6} - \frac{\pi}{6}\right) = \sin\left(\frac{5\pi}{6}\right)$$

Subsequently, the solution is $\frac{5\pi}{6}$.

Examples

---

1. Demonstrate the triangle that satisfies the following equation pictorially, please and thank you.

$$\sin^{-1}\left(\frac{3}{5}\right)$$

Solution:

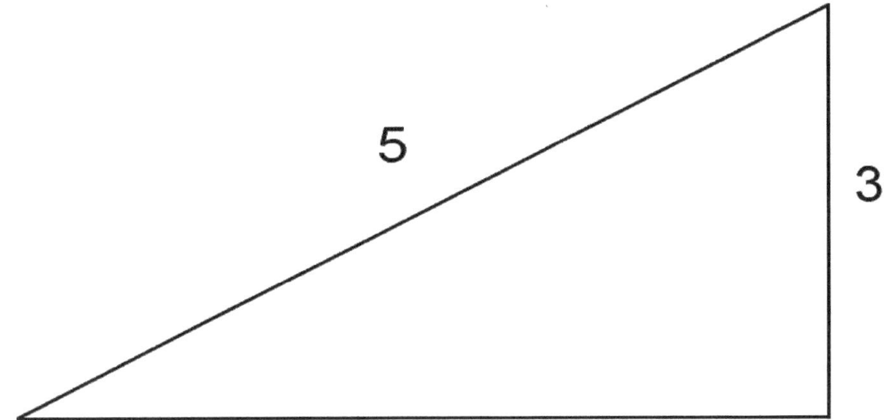

Figure 6.2.

2. Demonstrate the triangle that satisfies the following equation pictorially, please and thank you.

$$\sin^{-1}\left(\frac{5}{3}\right)$$

Solution:

Graphing such a triangle for real values is impossible. We have defined sin(x) as $\frac{Opposite}{Hypotenuse}$, and, as discussed earlier, the hypotenuse must be the largest side; therefore, graphing such a triangle is not achievable. Another way to illustrate the aforementioned point is as follows:

$\sin^{-1}(x) \mid x \in \mathbb{R}, -1 \leq x \leq 1\}.$

Please find the following angles with the assumption that arcsin(x) is a function rather than a relation.

3. $\arcsin\left(\frac{1}{2}\right)$

Solution:

We are interested to know the sine of what angle is $\frac{1}{2}$ in the first quadrant, which would be 30°or $\frac{\pi}{6}$.

---

4. $\arcsin(0)$

Solution:

We are interested to know the sine of what angle is 0 in the first quadrant, which would be 0°or 0 radians.

---

5. $\arcsin(-1)$

Solution:

We are interested to know the sine of what angle is -1 in the first quadrant, which would be -90°or $-\frac{\pi}{2}$ radians.

6.2. Inverse of cos(x)

As it was in the case for $\sin^{-1}(x)$, we can define $\cos^{-1}(x)$ also known as $\arccos(x)$ as follows:

$$\{\cos^{-1}(x) = y \mid x, y \in \mathbb{R}, 0 \leq y \leq \pi, -1 \leq x \leq 1\}$$

As we have discussed, the cosine of an angle is between -1 and 1. That is, the domain of arccosine is between -1 and 1. To get such values for the arccosine function, we can set the angle that can vary between 0 and π. That is, we can uniquely define the arccosine function's range given that $0 \leq y \leq \pi$. However, the arccosine relation can have a greater range as there can be more than one y value for every x. The following graph illustrates the arccosine function, and it should clarify why we have defined the domain and the range. Please note that the function does not have any asymptotes because it hits the function at the point $(-1, \pi)$ and $(1,0)$.

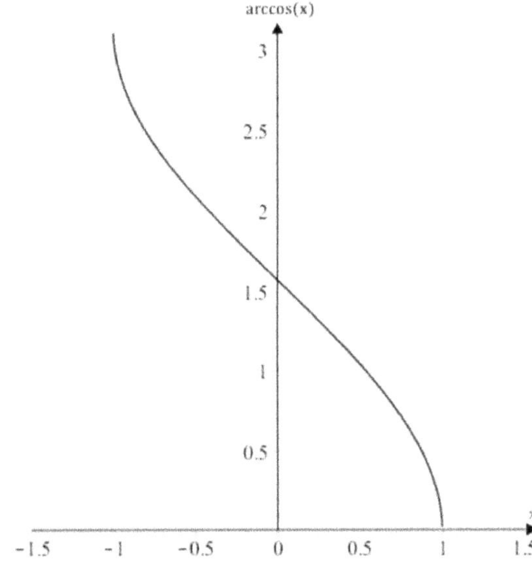

Figure 6.3.

Examples

6. Demonstrate the triangle that satisfies the following equation pictorially, please and thank you.

$$\cos^{-1}\left(\frac{1}{3}\right)$$

Solution:

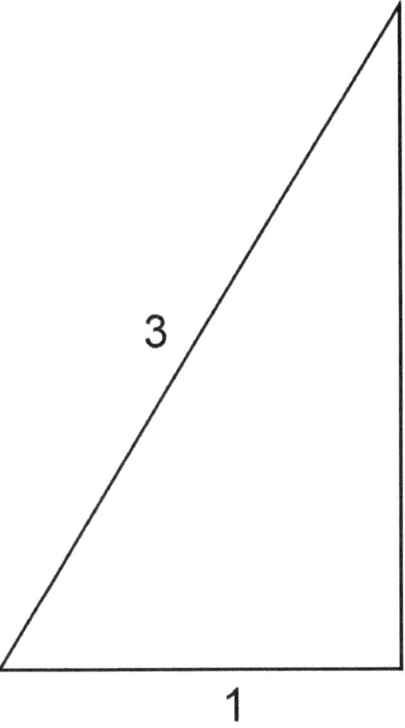

Figure 6.4.

Please find the following angles. You may assume arccos(x) is a relation rather than a function; therefore, please provide the complete range of solutions.

7. $\arccos\left(\frac{1}{2}\right)$, In the fourth quadrant

Solution:

We are trying to figure out the cosine of what angle is $\frac{1}{2}$ in a way that the angle lies in the fourth quadrant. We know $\cos(60^{\circ}) = \frac{1}{2}$, however, we are aiming for the answer in the 4th quadrant. Since we have the identity $\cos(\mathrm{x}) = \cos(-x)$, we can conclude $\cos(60^{\circ}) = \cos(-60^{\circ}) = \frac{1}{2}$. The -60° angle lies in the family of the solutions. The complete solution would be $-60^{\circ} + \mathrm{k}360^{\circ}$ or, in radian's form, it would be $-\frac{\pi}{6} + 2k\pi$, in which k is an integer.

---

8. $\arccos(1)$

Solution:

We need to determine the cosine of what angle is 1. We know $\cos(0^{\circ}) = 1$. We can, therefore, say 0° is in the family of the solutions. The complete solution would be $0^{\circ} + \mathrm{k}360^{\circ}$ or, in radian's form, it would be $0 + 2k\pi$, in which k is an integer.

---

9. $\arccos\left(-\frac{\sqrt{2}}{2}\right)$, In the second quadrant

Solution:

What we are going to figure out is the cosine of what angle is $-\frac{\sqrt{2}}{2}$ in a way that the angle lies in the second quadrant. We know $\cos\,(45^{\circ}) = \frac{\sqrt{2}}{2}$; however, we are asked for the $-\frac{\sqrt{2}}{2}$. Since we have the identity $-\cos(\mathrm{x}) = \cos(\pi - x)$, we can conclude that $-\cos(45^{\circ}) = \cos(180^{\circ} - 45^{\circ}) = \cos(135^{\circ}) = -\frac{\sqrt{2}}{2}$; therefore, the angle is 135° is in the family of the solutions. The complete solution would be $135^{\circ} + \mathrm{k}360^{\circ}$, or, in radian's form, it would be $\frac{3\pi}{4} + 2k\pi$, in which k is an integer.

---

10. $\arccos\left(-\frac{1}{2}\right)$, In the first quadrant

Solution:

Oops, cosine is not negative in the first quadrant. Who is writing these questions?

---

11. $\arccos\left(-\frac{3}{2}\right)$

Solution:

No solution for real values. Didn't we say the cosine of an angle is between -1 and 1 for the real numbers? We did :D. $-\frac{3}{2}$ is not in the range of cosine function (it is less than -1).

---

12. $\arccos\left(\frac{\sqrt{2}}{2}\right)$

Solution:

We are aiming to evaluate the cosine of an angle equal to $\frac{\sqrt{2}}{2}$. We know $\cos(45^{o}) = \frac{\sqrt{2}}{2}$; however, we are not given a particular quadrant. That is, we need to find all the possible solutions.

There are two families of solutions that we need to address, one being $\cos(45^{o}) = \frac{\sqrt{2}}{2}$.

The other can be obtained using identity $\cos(\mathrm{x}) = \cos(-x)$; we can then conclude $\cos(45^{o}) = \cos(-45^{o}) = \frac{\sqrt{2}}{2}$. Therefore, the angle is -45° is in the other family of the solutions.

The complete solution would be $45^{o} + \mathrm{k}360^{o}$, or, in radian's form, it would be $\frac{\pi}{4} + 2k\pi$ along with $-45^{o} + \mathrm{k}360^{o}$, or, in radian's form, it would be $-\frac{\pi}{4} + 2k\pi$, in which k is an integer.

---

13. $\arccos(0)$

Solution:

We want to figure out the cosine of what angle is 1. We know that $\cos(90^{o}) = 0$. Hence, we can identify 90° to be in the family of the solutions. The complete solution would be $90^{o} + \mathrm{k}180^{o}$, or, in radian's form, it would be $\frac{\pi}{2} + k\pi$, in which k is an integer.

Question, we know the period of cosine is 360° or 2π. Why did we add 180° or π instead of 360° or 2π to get the other solutions? The reason is that $\cos\ (90^{o}) = \cos(-90^{o}) = 0$. That is, we could have said the following:

$90^{o} + \mathrm{k}360^{o}$ and $-90^{o} + \mathrm{k}360^{o}$ for each family of solutions, respectively. Now let's see what the result would have been by writing a few of the angles:

$90^{o} + \mathrm{k}360^{o} = 90^{o}, 450^{o}, 810^{o}, \ldots$

$-90^{o} + \mathrm{k}360^{o} = -90^{o}, 270^{o}, 630^{o}, \ldots$

Now let's put the few we wrote in order.

$-90^{o}, 90^{o}, 270^{o}, 450\ ^{o}, 630^{o}, 810^{o}, \ldots$

As you can see in the combined version, the difference between each term is exactly 180°. The reason for that is the difference between the angles of $\cos\ (90^{o})$ and $\cos\ (-90^{o})$ is exactly 180°. That is why such an arrangement is possible.

14. $\arccos\left(\frac{\sqrt{3}}{2}\right)$, In the fourth quadrant

Solution:

We hope to figure out the cosine of what angle is $\frac{\sqrt{3}}{2}$ in a way that the angle lies in the fourth quadrant. We know cos $(30^o) = \frac{\sqrt{3}}{2}$; however, we are asked for the angle in the 4th quadrant. We utilize the identity $\cos(x) = \cos(-x)$ in order to get cos $(30^o) = \cos(-30^o) = \frac{\sqrt{3}}{2}$. Consequently, the angle -30° is in the family of the solutions. The complete solution would be $-30^o + k360^o$, or, in radian's form, it would be $\frac{\pi}{6} + 2k\pi$, in which k is an integer.

---

15. $\arccos(-1)$, in the third quadrant

Solution:

Our goal is to figure out the cosine of what angle is -1 in a way that the angle lies in the third quadrant. We know cos $(0^o) = 1$. To find the cosine of the angle that is equal to -1, we can use the following identity: $-\cos(x) = \cos(\pi - x)$ we can conclude $-\cos(0^o) = \cos(180^o - 0^o) = \cos(180^o) = -1$. We can, therefore, say that 180° is in the family of the solutions. The complete solution would be $180^o + k360^o$ or, in radian's form, it would be $\pi + 2k\pi$, in which k is an integer.

If you want to be very technical, you may argue the value of $\arccos(-1)$ does not fall in any quadrant since it is between the second and third quadrants. You are absolutely correct, good catch :D.

6.3. Inverse of tan(x)

As it was the case for $\sin^{-1}(x)$ and $\cos^{-1}(x)$, we can define $\tan^{-1}(x)$, also known as $\arctan(x)$, as follows:

$$\{ \tan^{-1}(x) = y \mid x, y \in \mathbb{R}, -\frac{\pi}{2} < y < \frac{\pi}{2}\}$$

As we have discussed, the tangent of an angle is anywhere from $-\infty$ to $+\infty$. That is, the domain of arctangent can take any real value. In order to get such values, the angle can vary between $-\frac{\pi}{2}$ and $\frac{\pi}{2}$ exclusively, which means it is not included the values $-\frac{\pi}{2}$ and $\frac{\pi}{2}$, the reason being, we have defined $\tan(x) = \frac{\sin(x)}{\cos(x)}$ denominator for such values would be zero. The arctangent function's range can be uniquely defined as $-\frac{\pi}{2} < y < \frac{\pi}{2}$. However, the arctangent relation can have a greater range as there can be more than one y value for every x. The following graph illustrates the graph of arctangent function, and it should clarify why we have defined domain and range as we have. Please note that the functions have two horizontal asymptotes at $y = -\frac{\pi}{2}$ and $y = \frac{\pi}{2}$.

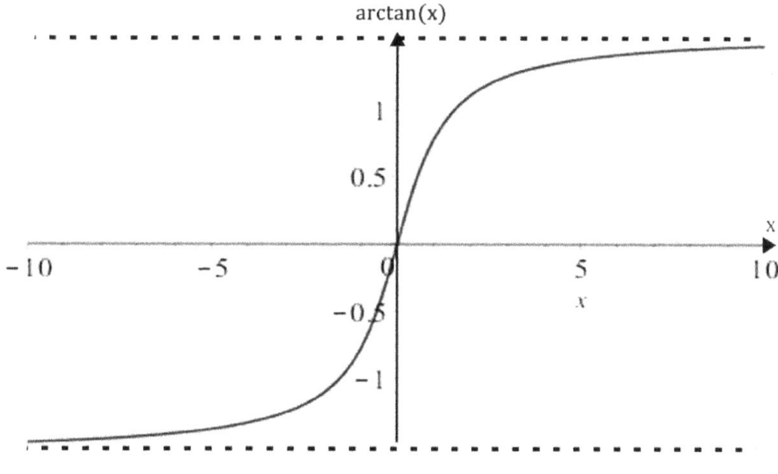

Figure 6.5.

Examples

16. Demonstrate the triangle that satisfies the following equation pictorially, please and thank you.

$$\tan^{-1}\left(\frac{3}{4}\right)$$

Solution:

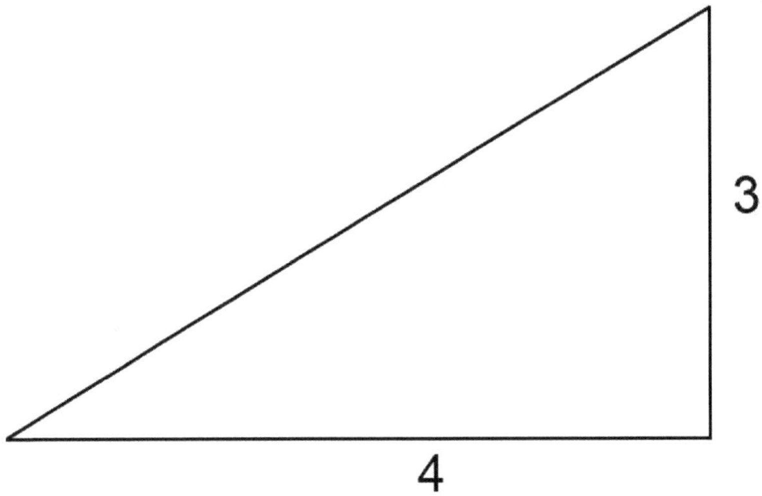

Figure 6.6.

Please find the following angles with the assumption that arctan(x) is a function rather than a relation.

17. $\arctan(0)$

Solution:

We are interested to know the tan of what angle is 0 in the domain of the arctan, which would be 0°or 0 radians. An easy way, other than looking at the table for tangent value, is to say sine what angle over cosine of the same angle would be 0, which happens to be 0°or 0 radians. Please consider the following calculation for a better understanding:

$$\frac{\sin(0^{o})}{\cos(0^{o})} = \frac{0}{1} = 0$$

---

18. $\arctan(1)$

Solution:

We are interested to know the tan of what angle is 1 in the domain of the arctan, which would be 45°or $\frac{\pi}{4}$ radians. An easy way, other than looking at the table for tangent value, is to say sine what angle over cosine of the same angle would be 1, which happens to be 45°or $\frac{\pi}{4}$. Please consider the following calculation for better understanding:

$$\frac{\sin(45^{o})}{\cos(45^{o})} = \frac{\frac{\sqrt{2}}{2}}{\frac{\sqrt{2}}{2}} = 1$$

---

19. $\arctan(\sqrt{3})$

Solution:

We are interested to know the tangent of what angle is $\sqrt{3}$ in the first quadrant, which would be 60°or $\frac{\pi}{3}$ radians. An easy way, other than looking at the table for tangent value, is to say sine what angle over cosine of the same angle would be $\sqrt{3}$, which happens to be

60°, or $\frac{\pi}{3}$. Please consider the following calculation for better understanding:

$$\frac{\sin(60^{o})}{\cos(60^{o})} = \frac{\frac{\sqrt{3}}{2}}{\frac{1}{2}} = \frac{\sqrt{3}}{2} \cdot \frac{2}{1} = \sqrt{3}$$

---

20. $\arctan(-1)$

Solution:

We are interested to know the tangent of what angle is -1 in the fourth quadrant, which would be-45°or $-\frac{\pi}{4}$ radians. An easy way, other than looking at the table for tangent value, is to say sine what angle over cosine of the same angle would be -1, which happens to be -45°, or $-\frac{\pi}{4}$. Please consider the following calculation for better understanding:

$$\frac{\sin(-45^{o})}{\cos(-45^{o})} = \frac{-\frac{\sqrt{2}}{2}}{\frac{\sqrt{2}}{2}} = -1$$

6.4. Inverse of cot(x)

As was the case in the aforementioned inverse trigonometric ratios, we can define $\cot^{-1}(x)$, also known as $\mathrm{arccot}(x)$, as follows:

$$\{\cot^{-1}(x) = y \mid x, y \in \mathbb{R}, 0 < y < \pi\}$$

As we have discussed, the cotangent of an angle is anywhere from $-\infty$ to $+\infty$. That is the domain of arccot(x) can take any real value. In order to get such values, the angle can vary between 0 and π exclusively, which means it does not include the values 0 and π, the reason being that we have defined the $\cot(x) = \frac{\cos(x)}{\sin(x)}$ denominator for such values would be zero. The arctangent function's range can be uniquely defined as $-\frac{\pi}{2} < y < \frac{\pi}{2}$. However, the arctangent relation can have a greater range as there can be more than one y value for every x. The following graph illustrated the graph of arctangent function, and it should clarify why we have defined domain and range as we did. Please note that the functions have two horizontal asymptotes at $y = 0$ and $y = \pi$.

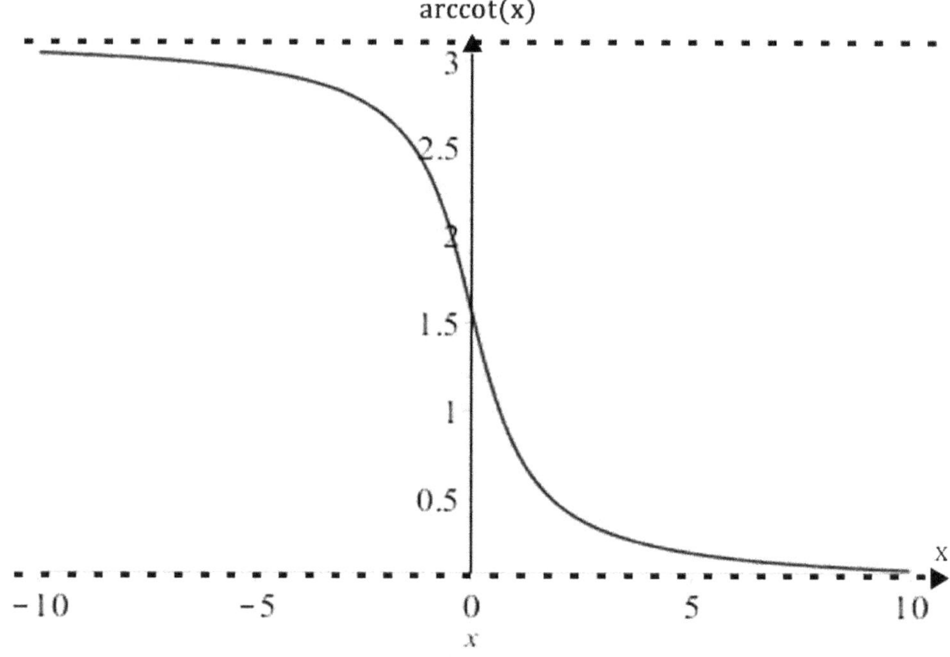

Figure 6.7.

Examples

---

21. Demonstrate the triangle that satisfies the following equation pictorially, please and thank you.

$$\cot^{-1}\left(\frac{8}{5}\right)$$

Solution:

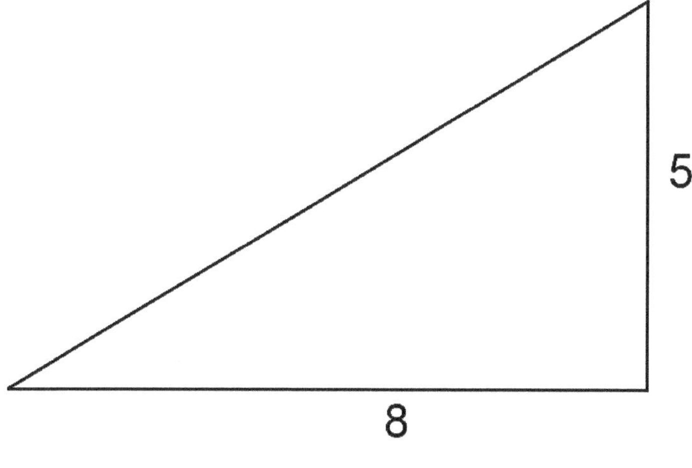

Figure 6.8.

Please find the following angles with the assumption that arccot(x) is a function rather than a relation.

22. $\text{arccot}(0)$

Solution:

We are interested to know the cotangent of what angle is 0 in the domain of the arccot, which would be 90°or $\frac{\pi}{2}$ radians. An easy way, other than looking at the table for cotangent value, is to say cosine what angle over sine of the same angle would be 0, which happens to be 90°or $\frac{\pi}{2}$ radians. Please consider the following calculation for better understanding:

$$\frac{\cos(90^{o})}{\sin(90^{o})} = \frac{0}{1} = 0$$

---

23. $\text{arccot}\left(\frac{\sqrt{3}}{3}\right)$

Solution:

We are interested to know the cotangent of what angle is 1 in the domain of the arccot, which would be 60°or $\frac{\pi}{3}$ radians. An easy way, other than looking at the table for cotangent value, is to say cosine what angle over sine of the same angle would be 1, which happens to be 60°or $\frac{\pi}{3}$. Please consider the following calculation for better understanding:

$$\frac{\cos(60^o)}{\sin(60^o)} = \frac{\frac{1}{2}}{\frac{\sqrt{3}}{2}} = \frac{1}{2}\frac{2}{\sqrt{3}} = \frac{1}{\sqrt{3}}$$

To rationalize the denominator, we can multiply the numerator and denominator by $\sqrt{3}$

$$\frac{1}{\sqrt{3}}\frac{\sqrt{3}}{\sqrt{3}} = \frac{\sqrt{3}}{3}$$

---

24. $\mathrm{arccot}(\infty)$

Solution:

Infinity is not in the domain; it is a concept, not a value.

Note, if you are looking to get the limit of $\operatorname{arccot}(x)$ as x approaches infinity, then it is a different story, and we can proceed as follows. We are interested to know the cotangent of what angle is ∞, which would be 0°or 0 radians. An easy way, other than looking at the table for tangent value, is to say cosine what angle over sine of the same angle would be ∞, which happens when the dominator is 0. In such a case, the fraction would be undefined. Since $\cot(x) = \frac{\cos(x)}{\sin(x)}$, we want the denominator to be zero while the numerator is a positive value to get ∞. We can achieve such a condition if the angle is 0°, or 0. Please consider the following calculation for better understanding:

$$\lim_{x \to 0^{\circ}} \left(\frac{cos(x)}{sin(x)} \right) = \frac{cos(0^{\circ})}{sin(0^{\circ})} = \frac{1}{0} = \infty$$

Briefly, what I mean by limit is I have an ice cream. You may look at it and get close to it. I mean very close, but you are not able to reach it. However, without reaching the ice cream, you can still tell what kind it is and what flavor it contains. The reason I don't let you get it is that the reality is the ice cream may or may not exist in the first place. In cases in which it exists, you are even able to touch it to get a better sense.

---

25. $\operatorname{arccot}(-1)$

Solution:

We are interested to know the cotangent of what angle is -1 in the fourth quadrant, which would be -45°or $-\frac{\pi}{4}$ radians. An easy way, other than looking at the table for cotangent value, is to say cosine what angle over sine of the same angle would be -1, which happens to be -45°,or $-\frac{\pi}{4}$.

Please consider the following calculation for better understanding:

$$\frac{\cos(-45^{\circ})}{\sin(-45^{\circ})} = \frac{\frac{\sqrt{2}}{2}}{-\frac{\sqrt{2}}{2}} = -1$$

Please note the angle must be between 0° and 180°. Also, the period for cot(x) is 180°. Therefore, the angle would be $-45^{\circ} + 180^{\circ} = 135^{\circ}$ or $-\frac{\pi}{4} + \pi = \frac{3\pi}{4}$.

6.5. Inverse of sec(x)

As was the case in the aforementioned inverse trigonometric ratios, we can define $\sec^{-1}(x)$ also known as $\mathrm{arcsec}(x)$, as follows:

$$\{ \sec^{-1}(x) = y \mid x, y \in \mathbb{R}, 0 \leq y < \frac{\pi}{2} \cup \frac{\pi}{2} \leq y \leq \pi , 1 \leq |x| \}$$

As we have discussed, the secant of an angle can accept any value greater than or equal to 1 or less than or equal to -1. That is, we can define the domain of arcsec(x) as $x \leq -1 \cup 1 \leq x$. Therefore, the angle can take all the real values except for $\frac{\pi}{2}$, the reason being that we have defined $\sec(x) = \frac{1}{\cos(x)}, \frac{\pi}{2}$ makes the denominator zero. We can uniquely identify the arcsec(x) function's range as $0 \leq y < \frac{\pi}{2} \cup \frac{\pi}{2} < y \leq \pi$. However, the arcsec(x) relation can have a greater range as there can be more than one y value for every x. The following graph illustrates the arcsec(x) function and clarifies how we have defined the domain and range of it. Please note the function has a horizontal asymptote at $y = \frac{\pi}{2}$. Also, there are no vertical asymptotes at $x = 1$ or $x = -1$ as the function would include the point $(1,0)$ and $(-1, \pi)$.

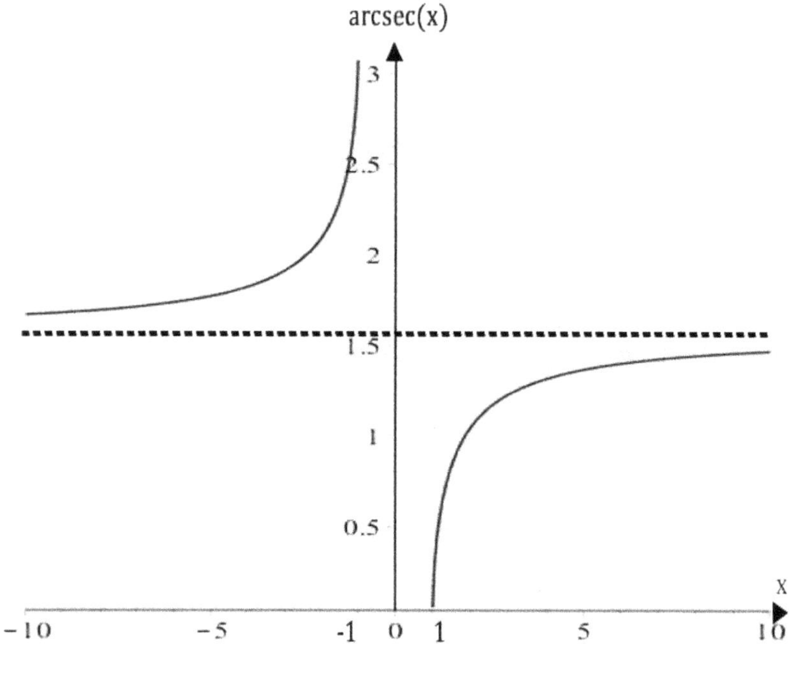

Figure 6.9.

Examples

26. Demonstrate the triangle that satisfies the following equation pictorially, please and thank you.

$$\sec^{-1}\left(\frac{5}{6}\right)$$

Solution:

It is not possible to draw such a triangle as the domain of arcsec is undefined for the aforementioned values. In other words, the hypotenuse cannot be less than the other sides.

---

27. Demonstrate the triangle that satisfies the following equation pictorially, please and thank you.

$$\sec^{-1}\left(\frac{5}{3}\right)$$

Solution:

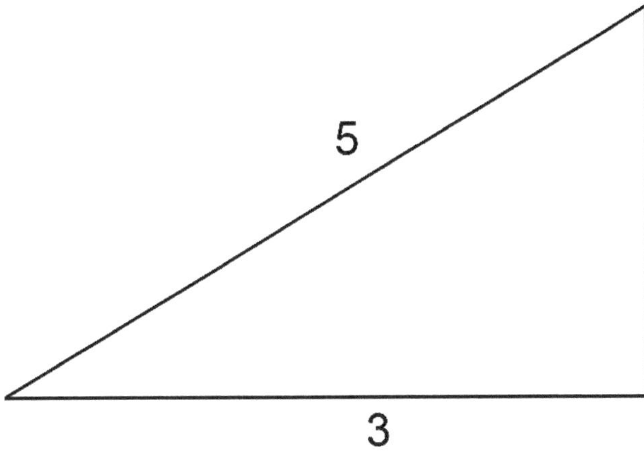

Figure 6.10.

Please find the following angles with the assumption that arcsec(x) is a function rather than a relation.

28. $\text{arcsec}(1)$

Solution:

We are interested to know the secant of what angle is 1 in the domain of the arcsec, which would be 0°or 0 radians. An easy way, other than looking at the table for secant value, is to say one over cosine of what angle is 1, which happens to be $0°$or 0 radians. Please consider the following calculation for better understanding:

$$\frac{1}{\cos(0°)} = \frac{1}{1} = 1$$

---

29. $\text{arcsec}(-\sqrt{2})$

Solution:

We are interested to know the secant of what angle is $-\sqrt{2}$ in the domain of the arcsec, which would be $135°$or $\frac{3\pi}{4}$ radians. An easy way, other than looking at the table for secant value, is to say one over cosine of what angle is $-\sqrt{2}$, which happens to be $135°$or $\frac{3\pi}{4}$ radians. Please consider the following calculation for better understanding:

$$\frac{1}{\cos(135°)} = \frac{1}{-\frac{\sqrt{2}}{2}} = 1\left(-\frac{2}{\sqrt{2}}\right) = -\frac{2}{\sqrt{2}}$$

To rationalize the denominator, we can multiply the numerator and denominator by $\sqrt{2}$

$$-\frac{2}{\sqrt{2}}\frac{\sqrt{2}}{\sqrt{2}} = \frac{-2\sqrt{2}}{2} = -\sqrt{2}$$

---

30. $\text{arcsec}\left(\frac{2}{\sqrt{3}}\right)$

Solution:

We are interested to know the secant of what angle is $\frac{2}{\sqrt{3}}$ in the domain of the arcsec, which would be 30°or $\frac{\pi}{6}$ radians. An easy way, other than looking at the table for secant value, is to say one over cosine of what angle is $\frac{2}{\sqrt{3}}$, which happens to be 30° or $\frac{\pi}{6}$ radians. Please consider the following calculation for better understanding.

$$\frac{1}{\cos(30^{o})} = \frac{1}{\frac{\sqrt{3}}{2}} = 1\left(\frac{2}{\sqrt{3}}\right) = \frac{2}{\sqrt{3}}$$

6.6. Inverse of csc(x)

As was the case in the aforementioned inverse trigonometric ratios, we can define $\csc^{-1}(x)$, also known as $\text{arccsc}(x)$, as follows:

$$\{\ \csc^{-1}(x) = y \mid x, y \in \mathbb{R}, -\frac{\pi}{2} \leq y < 0\ \cup 0 \leq y \leq \frac{\pi}{2}, 1 \leq |x|\ \}$$

As we have discussed, the secant of an angle can accept any value greater than or equal to 1 or less than or equal to -1. That is, we can define the domain of arcsec(x) as $x \leq -1\ \cup 1 \leq x$. Therefore, the angle can take all the real values except for 0, the reason being that we have defined $\csc(x) = \frac{1}{\sin(x)}$, 0 makes the denominator zero. We can uniquely identify the arcsec(x) function's range as:

$$-\frac{\pi}{2} \leq y < 0\ \cup 0 < y \leq \frac{\pi}{2}.$$

However, the arcsec(x) relation can have a greater range as there can be more than one y value for every x. The following graph illustrates the arcsec(x) function and clarifies how we have defined the domain and range of it.

Please note the function has a horizontal asymptote at $y = 0$. Also, there are no vertical asymptotes at $x = 1$ or $x = -1$ as the function would include the point $\left(-1, -\frac{\pi}{2}\right)$ and $\left(1, \frac{\pi}{2}\right)$.

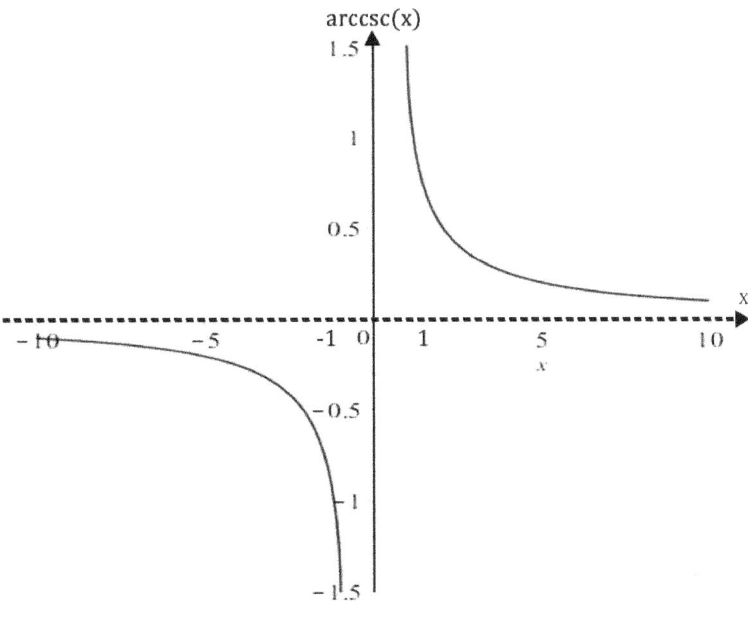

Figure 6.11.

Examples

31. Demonstrate the triangle that satisfies the following equation pictorially, please and thank you.

$$\csc^{-1}\left(\frac{11}{4}\right)$$

Solution:

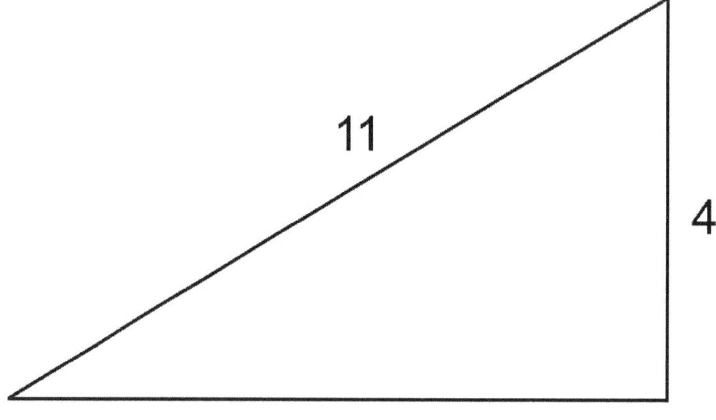

Figure 6.12.

Please find the following angles with the assumption that arccsc(x) is a function rather than a relation.

32. $\text{arccsc}(1)$

Solution:

We are interested to know the cosecant of what angle is 1 in the domain of the arccsc, which would be 90^oor $\frac{\pi}{2}$ radians. An easy way, other than looking at the table for cosecant value, is to say one over sine of what angle is 1, which happens to be 90^oor $\frac{\pi}{2}$ radians. Please consider the following calculation for better understanding:

$$\frac{1}{\sin(90^o)} = \frac{1}{1} = 1$$

---

33. $\text{arccsc}(\sqrt{2})$

Solution:

We are interested to know the cosecant of what angle is $\sqrt{2}$ in the domain of the arccsc, which would be $45°$or $\frac{\pi}{4}$ radians. An easy way, other than looking at the table for cosecant value, is to say one over sine of what angle is $\sqrt{2}$, which happens to be $45°$or $\frac{\pi}{4}$ radians. Please consider the following calculation for better understanding:

$$\frac{1}{\sin(45°)} = \frac{1}{\frac{\sqrt{2}}{2}} = 1\left(\frac{2}{\sqrt{2}}\right) = \frac{2}{\sqrt{2}}$$

To rationalize the denominator, we can multiply the numerator and denominator by $\sqrt{2}$

$$\frac{2}{\sqrt{2}}\frac{\sqrt{2}}{\sqrt{2}} = \frac{2\sqrt{2}}{2} = \sqrt{2}$$

---

34. $\text{arccsc}\left(-\frac{2}{\sqrt{3}}\right)$

Solution:

We are interested to know the cosecant of what angle is $-\frac{2}{\sqrt{3}}$ in the domain of the arccsc, which would be$-60°$or $-\frac{\pi}{3}$ radians. An easy way, other than looking at the table for cosecant value, is to say one over sine of what angle is$-\frac{2}{\sqrt{3}}$, which happens to be$-60°$or $-\frac{\pi}{3}$ radians. Please consider the following calculation for better understanding:

$$\frac{1}{\sin(-60^o)} = \frac{1}{-\frac{2}{\sqrt{3}}} = 1\left(-\frac{\sqrt{3}}{2}\right) = -\frac{\sqrt{3}}{2}$$

---

35. $\text{arccsc}\left(-\frac{1}{2}\right)$

Solution:

We are interested to know the cosecant of what angle is $-\frac{1}{2}$ if such value is not in the arccsc's domain. That is, one over sine of an angle is $-\frac{1}{2}$. In other words, we have the following.

$$\frac{1}{\sin(x)} = -\frac{1}{2} => \sin(x) = -2$$

As we have defined, the sine of real numbers is between -1 and 1. Therefore, the value does not exist.

Index:

www.ingramcontent.com/pod-product-compliance
Lightning Source LLC
Chambersburg PA
CBHW061616210326
41520CB00041B/7468

* 9 7 8 0 9 9 4 0 7 3 9 4 5 *